MW01076566

Earth History

Developed at
The Lawrence Hall of Science,
University of California, Berkeley
Published and distributed by
Delta Education,
a member of the School Specialty Family

1558514
978-1-62571-785-6
Printing 1 –3/2017
Webcrafters, Madison, WI

Table of Contents

Readings

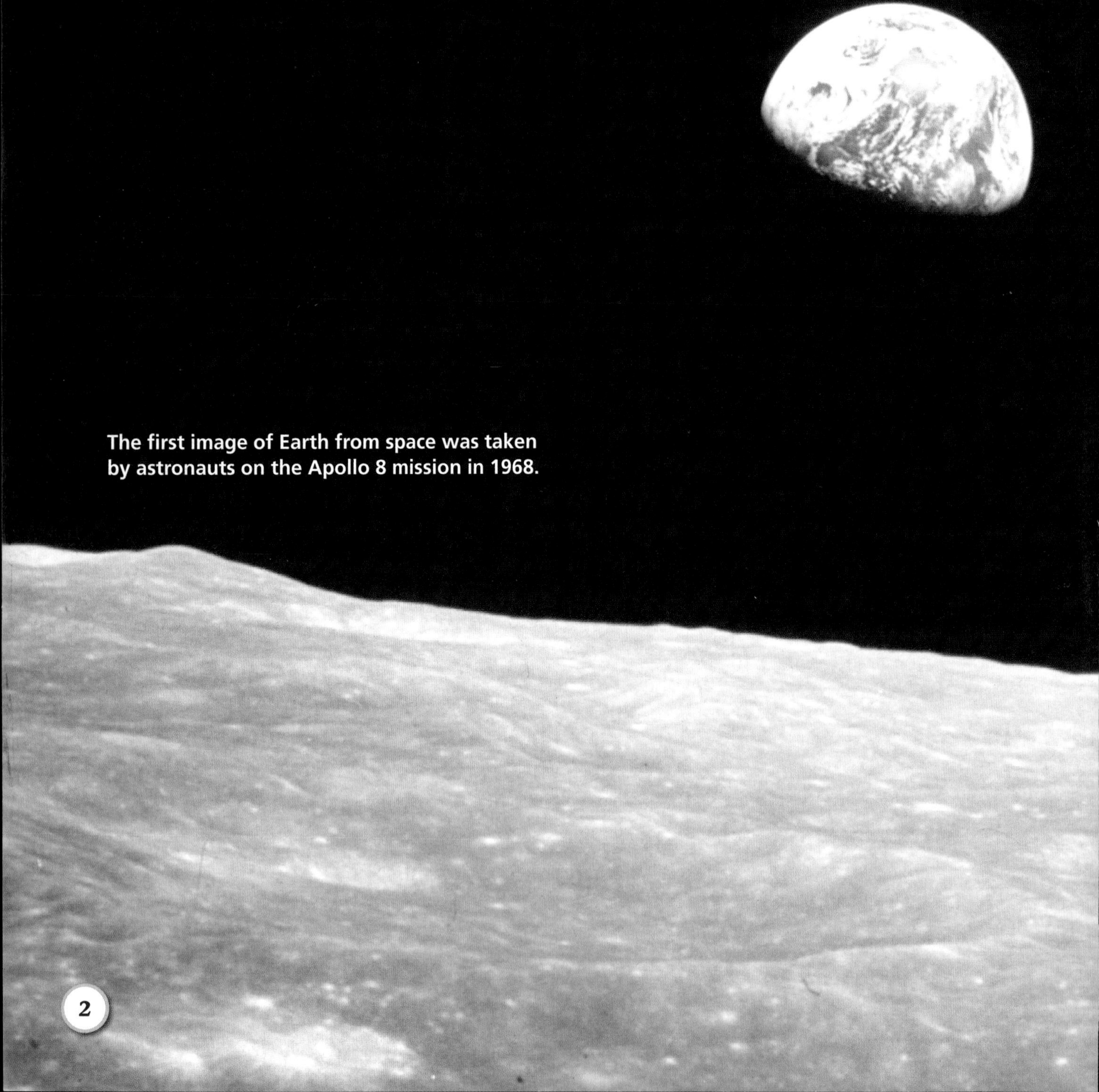

The first image of Earth from space was taken by astronauts on the Apollo 8 mission in 1968.

Seeing Earth

Today, we have all sorts of tools for viewing Earth. Anyone who has access to the Internet can see Earth from above, using tools like Google Earth™.

With the tap of a finger, you can fly from Portugal to the Amazon, explore the ocean floor, and fly thousands of kilometers into space.

Earth History Resources

The first **geologists** (scientists who study Earth) did not use **satellite** images to understand this planet. They relied on their experiences in the field, observing **landforms** and rocks. When you hold a rock, you hold an actual piece of Earth. You can observe it with your senses.

As modern-day geologists, we will use multiple resources to begin to understand the story of Earth. Our resources include satellite images and **models**, as well as actual field samples.

Geologists use many sophisticated tools and models—and their senses—to study the rocks that make up Earth.

Image Sources

Where do Google Earth™ images come from? They do not all come from one space camera that sends real-time images of Earth. For example, you cannot zoom in on your house and then walk outside to see yourself. The images come from different sources, including airplanes, satellites, drones, even kites. Many of the images are not photographs, but are three-dimensional computer-generated representations of Earth's **terrain**.

You use your computer to travel down the Colorado River and see the high walls of the Grand Canyon. You are not looking at a single image, taken from space. The image combines photos from above with data about the terrain. The result is like a giant jigsaw puzzle or mosaic. Many images stitched together are used to represent Earth.

This aerial view of the Grand Canyon, pieced together from several images, shows the rugged terrain from 17 km high. Satellite imagery and technology allow us to explore Earth in amazing new ways.

GeoEye-1

Most of the images in Google Earth™ come from satellites that are flying hundreds of kilometers above Earth. Thousands of satellites orbit the planet. A few have the single task of observing and recording Earth's surface. These **Earth-imaging satellites** cannot capture the entire Earth with one shot. In 1 day, a typical satellite can image less than 1 percent of Earth's surface.

The satellite that provides the most detailed images for Google Earth™ is called GeoEye-1. It records images of things as small as 60 centimeters (cm) wide. Using GeoEye-1's camera, you could see a brightly colored newspaper lying in the street, but you could not read the headlines. GeoEye-1 orbits at 677 kilometers (km) from Earth. That is 60 times higher than the flight of a typical passenger plane.

Did You Know?

When GeoEye-1 was launched in 2008, it was the highest resolution commercial Earth-imaging satellite.

GeoEye-1, launched in 2008, can produce satellite imagery of more than 350,000 square km of Earth's surface every day.

This image of Mount St. Helens was taken by a NASA satellite that can detect infrared radiation to measure heat. The areas in red indicate recent lava activity at the volcano that erupted violently in 1980.

Satellites like GeoEye-1 do not work like a regular digital camera. The GeoEye-1 camera can detect light that people cannot see. For example, satellites can detect hot areas where rain forest is being cleared and burned. They can also detect the temperature of ocean currents.

A satellite image gets processed into something that we can recognize as Earth's surface. After processing, the image can become a piece of the Google Earth™ mosaic. Not all images from GeoEye-1 are used in Google Earth™. It is hard to capture the perfect image with little shadow and no clouds. Sometimes the Google Earth™ image has lines where two satellite images were patched together. The finished mosaic might include satellite images that are a few months old and images that are several years old.

Think Questions

1. **How does Google Earth™ get its images?**
2. **How do tools like Google Earth™ help geologists?**

Powell's Grand Canyon Expedition, 1869

Imagine setting out to explore the Grand Canyon without any idea about the land or water. Major John Wesley Powell did this very thing. Only five of the nine expedition members completed the journey.

As a young man, Powell (1834–1902) enlisted as a private in the Union Army at the start of the Civil War. He earned his rank as major during that war. He lost his right arm during the Battle of Shiloh in 1862.

After the war, Powell taught geology at Illinois Wesleyan University. In 1868, at the age of 34, he decided to explore and map the Colorado River and Grand Canyon region. This area of the country was a blank spot on all maps at the time. It was difficult to get funding for this trip, as many people thought it was impossible. With the money he collected, he designed four wooden boats. The boats had special watertight compartments to keep the supplies dry. Powell tied a chair in the middle of the smallest boat. From there, he could see over the rapids and warn the other boats of hazards.

Powell and his nine recruits are about to embark in four specially built boats on a daring exploratory journey into uncharted canyons.

About 6 weeks Into the journey, the expedition entered (and named) Desolation Canyon. The Green River, flowing in long, wide meanders, picked up speed and became rougher between steep stone walls.

The expedition had enough supplies to last 10 months. Powell and his team began their trip on the Green River at Green River City, Utah, on May 24, 1869. Powell kept a daily journal on long strips of brown paper. He made the strips into little books bound with leather. His journals included both scientific data and geographic observations. He included notes about the men on the team and about his more risky adventures. He also included drawings of the rocks and landforms. There was no photographer on the expedition.

Four of Powell's team did not make it to the end of the trip. One decided he had had enough adventure before they even reached the Grand Canyon and left the river. Three others left near the end of the voyage when they decided to take their chances hiking overland out of the Grand Canyon. They were never heard from again.

Many thought Powell and all his team had died on the journey. But Powell and the remaining members reached the Grand Wash Cliffs on August 29, 1869. They had traveled more than 1,600 kilometers (km) on the river. Powell became a hero when he returned to the East.

His journal was published and became very popular. Powell's journal is still read today. It is called *The Exploration of the Colorado River and Its Canyons*. His 1869 expedition was as big an adventure as the first time men landed on the Moon in 1969. People marveled at his and his team's bravery and their stories of the Grand Canyon.

The following passages are adapted from *The Exploration of the Colorado River and Its Canyons*.

Later in life Powell directed the US Geological Survey and ran the Smithsonian's Bureau of Ethnology.

ꕥ May 24, 1869 ꕥ

The good people of Green River City turn out to see us start. We raise our little flag, push the boats from shore, and the swift current carries us down.

We have four boats. Three are built from oak and are 21 feet (6.4 meters [m]) long. They are sturdy boats with double ribs and double stern and sternpost. The bulkheads are also strengthened; they are divided into three compartments. Two of the compartments are decked and form watertight cabins. We hope that these compartments will help buoy the boats should waves roll over them in rough water. Four men can carry the empty boats easily.

The fourth boat is made of pine. It is 16 feet (4.9 m) long and has a sharp cutwater. It is built for fast rowing and divided into compartments like the others.

We take rations enough for 10 months. We take abundant supplies of clothing. We expect that we will layover somewhere when winter comes and the river is filled with ice. We also have a large quantity of ammunition and two or three dozen traps. For building cabins, repairing boats, and meeting other presently unknown needs, we have axes, hammers, saws, augers, and other tools, plus a large supply of nails and screws. For our scientific work, we have two sextants, four chronometers, a number of barometers, thermometers, compasses, and other instruments.

The Crew of the 1869 Expedition

J. C. Sumner
Civil War soldier, amateur hunter, mountain man

William H. Dunn
Hunter, trapper, mule packer

W. H. Powell
Civil War officer, John Wesley Powell's younger brother

G. Y. Bradley
Civil War officer

O. G. Howland
Printer by trade, editor by profession, hunter by choice

Seneca Howland
O. G. Howland's younger brother

Frank Goodman
British adventurer

W. R. Hawkins (aka Billy)
Civil War soldier, teamster, hunter, athlete, cook

Andrew Hall
Mule driver, boatman; at 19, the youngest member of the crew

℘ June 18, 1869 ℘

Bradley and I decide to climb Echo Rock. We start up a small canyon, then pass along a shelf of the wall. We climb 600 to 800 feet (183–244 m) when we meet a sheer precipice. We continue up. I go ahead with the barometer. We nearly reach the summit.

Supplies

General

Clothing for summer and winter
Ammunition
Traps (2 or 3 dozen)

Tools

Axes
Hammers
Saws
Augers
Other tools
A large supply of nails and screws

Scientific Instruments

Sextants (2)
Chronometers (4)
Barometers
Thermometers
Compasses
Other instruments

Food

They took enough rations for 10 months and hoped to supplement their diet with fresh meat and fish along the way.

Flour
Sugar
Coffee
Beans
Bacon
Other food

Boats

Boats
Kitty Clyde's Sister, No Name, and *Maid of the Canyon*
- Built from oak
- 6.4 m long
- Sturdy boats with double ribs, double stern, and sternpost
- Bulkheads strengthened and divided into three compartments
- Had two watertight compartments

Emma Dean—named after Powell's wife
- Made of pine
- 4.9 m long
- Built for fast rowing
- Had three compartments

I jump and gain a foothold in a small crevice, grasping another rock with my left hand. I can neither go up or down. I dare not let go.

I call Bradley for help. He finds a way to climb around me to a rock overhead, but he cannot reach me. Standing on my toes, my muscles begin to tremble. It is 60 to 80 feet (18–24 m) to the foot of the cliff. If I lose my hold I shall fall to the bottom and then perhaps roll over the bench and tumble farther.

At this point Bradley thinks to remove his pants, which he swings down to me. I hug close to the rock, let go of my hand, and seize the dangling legs. With Bradley's help I climb safely to the top.

℘ August 5, 1869 ℘

With some feeling of anxiety we enter a new canyon this morning. We have learned to observe closely the texture of the rock. In softer strata we have a quiet river, in harder we find rapids and falls. Below us are the **limestones** and hard **sandstones**, which we found in Cataract Canyon. This bodes toil and danger!

ꟿ August 13, 1869 ꟿ

We are now ready to start on our way down the Great Unknown. Our boats, tied to a common stake, chafe each other as they are tossed by the fretful river. They ride high and buoyant, for their loads are lighter than we could desire. We have but a month's rations remaining. The flour has been resifted through the mosquito-net sieve; the spoiled bacon has been dried and the worst of it boiled; the few pounds of dried apples have been spread in the sun and reshrunken to their normal bulk. The sugar has all melted and gone on its way down the river. But we have a large sack of coffee. The lightening of the boats has this advantage: they will ride the waves better and we shall have but little to carry when we make a portage. . . .

We have an unknown distance yet to run, an unknown river to explore. What falls there are, we know not; what rocks beset the channel, we know not; what walls rise over the river, we know not. Ah, well! We may conjecture many things. The men talk as cheerfully as ever; jests are bandied about freely this morning; but to me the cheer is somber and the jests are ghastly.

With some eagerness and some anxiety and some misgiving we enter the canyon below.

August 14, 1869

At daybreak we walk down the bank of the river, on a little sandy beach, to take a view of a new feature in the canyon. Heretofore hard rocks have given us bad river; soft rocks, smooth water; and a series of rocks harder than any we have experienced sets in. The river enters hard rock called gneiss! We can see but a little way into the granite gorge, but it looks threatening. . . .

Powell's second expedition, in 1871, produced what the first had not: a map, scientific publications, and photographs.

The walls now are more than a mile in height—a vertical distance difficult to appreciate. Stand on the south steps of the Treasury building in Washington and look down Pennsylvania Avenue to the Capitol; measure this distance overhead, and imagine cliffs to extend to that altitude, and you will understand what is meant. . . .

We camp for the night. It is raining hard, and we have no shelter, but find a few sticks, which have lodged in the rocks, and kindle a fire and have supper. We sit on the rocks all night, wrapped in our ponchos, getting what sleep we can.

Think Questions

1. **Pick one scene described by Powell in this section and draw a picture to show how it might have looked.**
2. **Describe several emotions Powell was probably feeling on the days described in this portion of his journal. What is your evidence he was experiencing these emotions?**

Grand Canyon Flood!

The scientists have placed their instruments, and the countdown is on. At just the right moment, an engineer opens the gates. Millions of liters of water pours from the dam, flooding the Colorado River in a giant experiment.

For thousands of years, **floods** in the Colorado River weathered layers of rock in the Colorado Plateau, carving the Grand Canyon. When humans built Glen Canyon Dam, it changed the natural flow of the Colorado River. Natural river levels vary, depending on rain and snow levels and severe storms. But a dam controls the water flow. It produces a fairly steady moderate flow into the Grand Canyon. As a result, the river's **ecosystem** changed, affecting animals and plants in the canyon.

Scientists monitoring these changes decided to create an experimental flood. Studying the flood might help them figure out a better way to manage the river's resources. They planned to let water pour out of Glen Canyon Dam as fast as possible for 7 days. This first flood started on March 26, 1996. The Grand Canyon became like a gigantic stream table for the scientists to study.

The gigantic Glen Canyon Dam was designed to harness the Colorado River to provide the water and power needs of millions of westerners. There were unintended consequences, however.

Why Create a Flood?

Before the dam was finished in 1963, the natural cycle of the river created an environment that supported river life. Each spring, melting snow upriver increased the flow of the river. More than 2.5 million liters (L) of water could roar past any point on the river every second. That's enough water to fill 40 backyard swimming pools every second! The raging water washed large quantities of **sediment** from the river bottom. Spring floods cleared vegetation from the sandbars and removed deposits of cobbles and boulders, which could clog the river or create rapids. During the summer months, less water flowed in the river, so sediments settled in sandbars and other places along the channel.

Glen Canyon Dam slowed the flow of water and cut down on flooding. Several unexpected things happened as a result.

- Sandbars got smaller.
- Vegetation began to grow into the river channel.
- Deposits of cobbles and boulders built up.
- Habitat for native fish decreased.

Scientists realized that floods in the Colorado River help keep sand in the canyon. They **inferred** that floods provide energy that moves the sand from the river bottom. This sand was deposited in sandbars high up on the riverbanks. There, sand was likely to erode away. Quiet water behind the sandbars create habitats for the native fish.

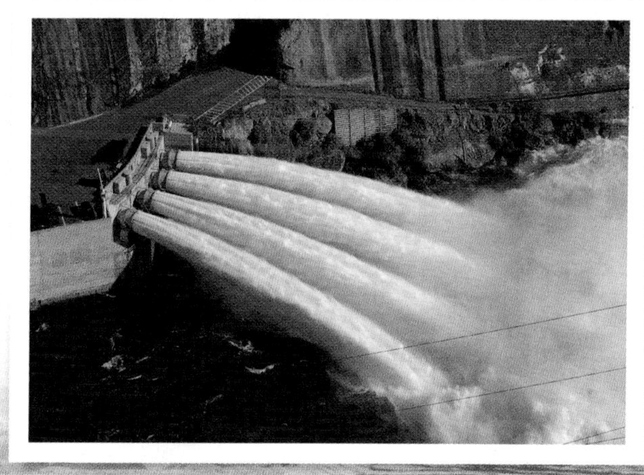

Since 1996, dam operators have conducted several high-flow releases of water to mimic natural flooding. The four huge jet outlet tubes bypass the power generators and do not produce electricity.

Scientists wanted to confirm their ideas about flood water and sandbars. They proposed releasing extra water from Glen Canyon Dam. Could they protect the disappearing sandbars? They planned a flood to find out.

Planning the Flood Experiment

The rate of water flowing from Glen Canyon Dam during the 7-day flood was 1,274,300 L per second, nearly the maximum possible. The plan was to collect data before, during, and after the flood. Geologists from the US Geological Survey and universities collected the data.

Water flow. How long would it take the water to flow from one place to another? Scientists monitored water flow at five stations along the river. They put harmless red dye in the river, and recorded how long it took the dye to reach different stations. They measured the height of the river at 40 sites. By combining water speed and height measurements, scientists could predict when the flood would arrive at any location along the river.

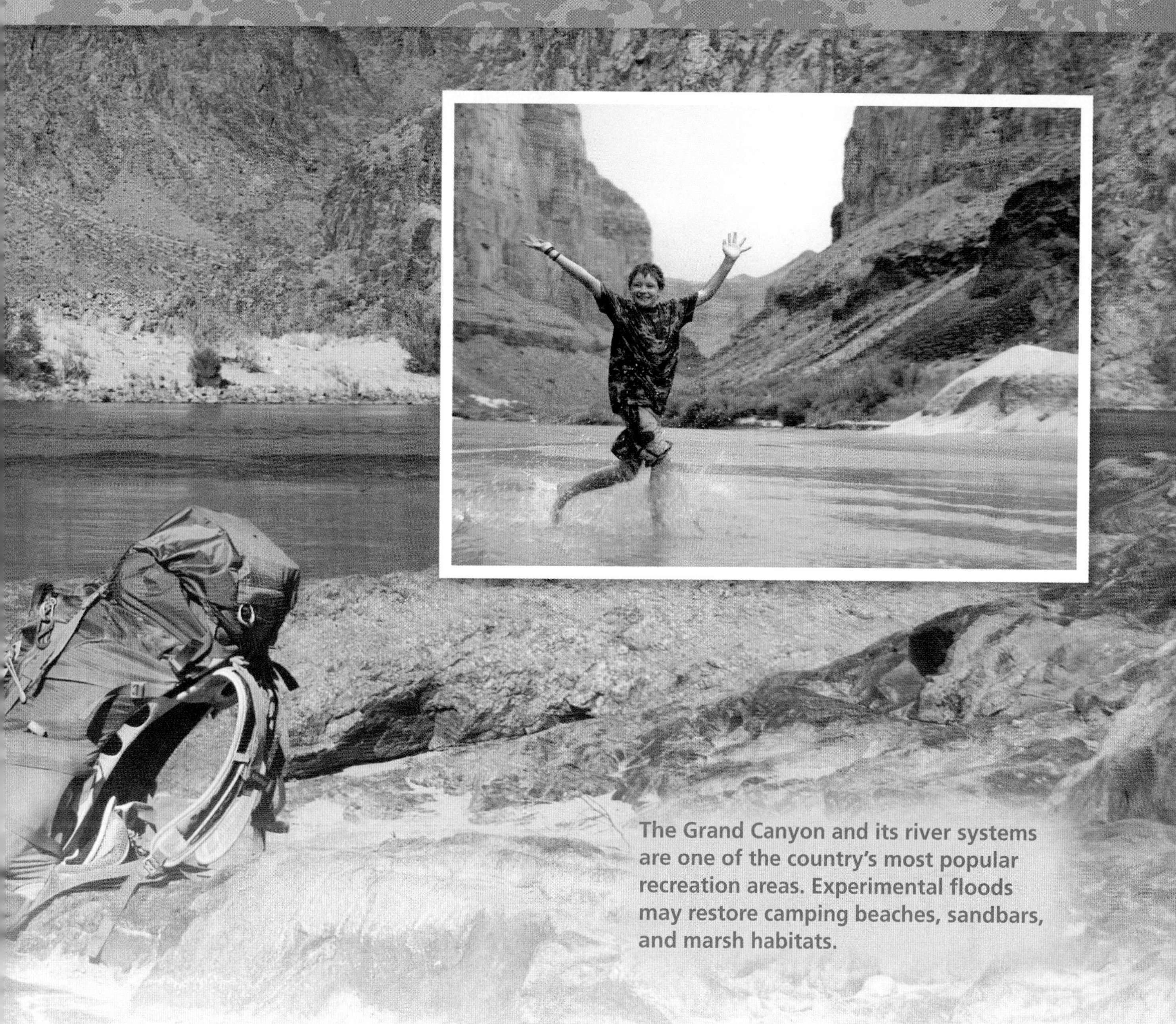

The Grand Canyon and its river systems are one of the country's most popular recreation areas. Experimental floods may restore camping beaches, sandbars, and marsh habitats.

Sediment transport. What would happen to beaches and sandbars during and after the flood? Would they be washed away, or would sand on the river bottom be deposited as new sandbars and beaches? Before the flood, scientists figured out how much sand was at 40 sites along the river. Sand can be stored in the riverbed under water. It can settle above the water level as beaches and sandbars. Boulders and cobbles deposit and accumulate wherever a side stream enters the main river. These deposits can dam the river and create rapids. Scientists wondered if a flood would pile up more material or wash the material downstream. They drilled holes into 150 basalt rocks, ranging from the size of basketballs to kitchen stoves. They put radio tags in each rock to track its movement. Other boulders were attached to cables and placed in the path of the flood. Scientists measured the amount of force the water put on these rocks.

The population of humpback chub declined when water temperatures decreased after the dam was built. This native species is now endangered.

River ecology. How would the flood affect river ecology? When the dam was completed and slowed the flow of water, the size and number of protective areas behind sandbars steadily declined. These habitat changes affected the native fish and plants along the river, a great concern to river biologists.

What the Flood Revealed

Here are some of the scientists' findings.

Water flow. The red dye showed that as the water rose and flooded the land, it slowed down in many places. The water moved even slower than when its levels were lower.

Some scientists thought that the flood might wash away some fish, snails, and marsh plants. But that did not happen. For example, one type of endangered snail survived very well, and scientists were able to determine the size of the snail population and what kinds of food the snails eat.

Five weeks after the flood, scientists studied Anasazi Marsh (Mile 43). They found that the limbs of willow trees protected the area during the flood. The marsh had little flood damage.

The first major whitewater downstream from Glen Canyon Dam is Badger Rapids. Imagine what the rapids look like during a flood.

Sediment transport. Observations showed that sediments moved differently during floods. Some evidence came from a study at Badger Rapids (Mile 8), the closest rapid to Glen Canyon Dam. There is little sediment **deposition** at Badger Rapids, so beach **erosion** is a constant threat. Five weeks after the flood, the beaches near Badger Rapids were wider than before the flood. Scientists inferred that the flood water brought up sand from the bottom of the river there. Most of it was pushed toward the sides of the river and settled on those beaches.

Scientists had placed the boulders with radio transmitters and force meters in the river at Lava Falls (Mile 179). They discovered that boulders the size of watermelons and televisions tumbled along the riverbed during the flood. Rocks of this size could easily break off pieces of **bedrock** in the river bottom. They could crush smaller rocks into pebbles and sand. The amount of energy needed to move boulders weighing a ton or more occurs only during a flood.

National Canyon (Mile 166) was another study site. Scientists collected water samples to observe the amount of sand in the water. They tracked how much sand was carried downstream and how much was deposited as sandbars and beaches along the river. Only about 10 percent of the sand settled out. The rest of the sand suspended in the flood water moved down the river.

River ecology. Native fish populations were not harmed by the flooding. Some of the larger fish might have been washed downstream. The small fish were safe in the natural eddies of the river. They stayed in these eddies after the water level went down. The flood water created more backwaters for these fish to spawn and grow in healthy numbers. The endangered snails survived in backwaters and eddies. The marshes withstood the flood waters and later provided shelter and food for the animals living in the river.

Conclusions

Scientists were able to collect and analyze a lot of data before, during, and after the flood. They ran similar experiments in 2004 and 2008 to collect even more data. They learned about the effects of the floods, and they continue to study the results. Their studies will help decide how to manage Glen Canyon Dam and the Colorado River in the future. As scientists continue to monitor erosion, deposition, and the canyon's ecology, they may someday create new experiments to answer new questions.

Before the flood, one National Canyon sandbar looked like this.

During the flood, the sandbar was completed submerged.

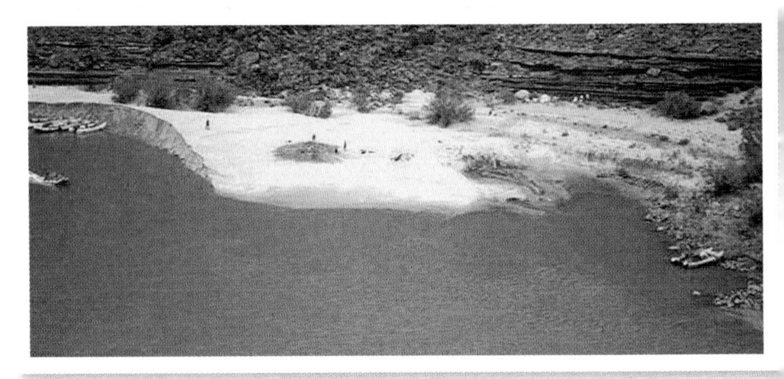

After the flood, the sand was much deeper and more widespread.

Take Note

Compare the sandbar before and after the flood. What evidence do you see that the sand is deeper?

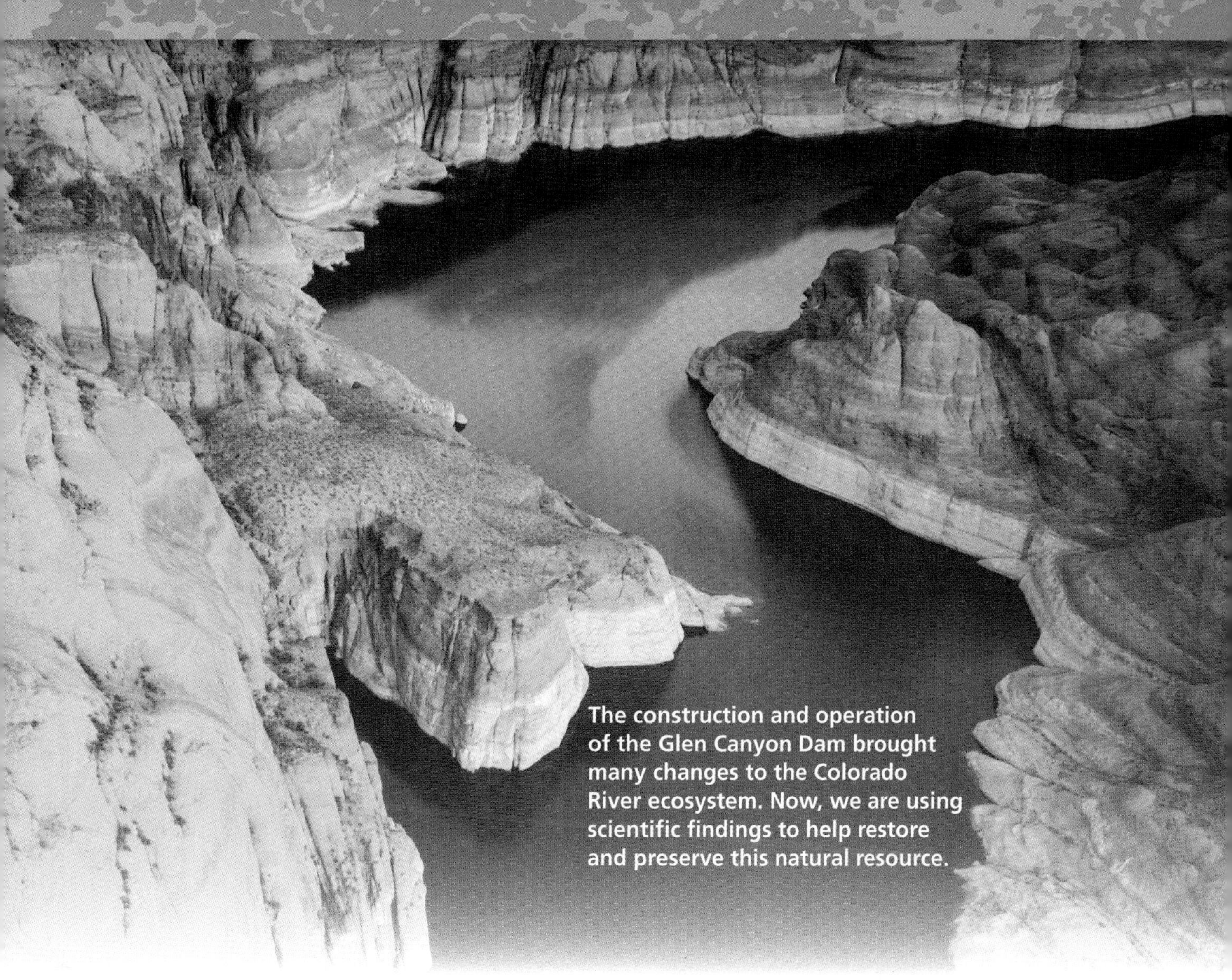

The construction and operation of the Glen Canyon Dam brought many changes to the Colorado River ecosystem. Now, we are using scientific findings to help restore and preserve this natural resource.

This article was adapted from material from the US Geological Survey. Photographs are from the US Geological Survey and the Sandbar Studies Lab, Department of Geology, Northern Arizona University, Flagstaff, Arizona.

For more information about studies at the Grand Canyon and Glen Canyon Dam, visit FOSSweb.com.

Think Questions

1. **How are the Grand Canyon flood investigations similar to the stream-table investigations you did in class and viewed in the online activity? How are they different?**
2. **Describe how you could set up a model stream table to demonstrate the effects of a flood.**
3. **Observe the photographs from National Canyon. Describe the changes you see before, during, and after the flood.**

Interesting rock formations like arches and pillars in Utah's Bryce Canyon were created by the natural forces of weathering.

Weathering and Erosion

What do sheer cliffs, balancing rocks, massive caves, and giant sand dunes have in common? They result from the processes of weathering and erosion. The same processes form and take away the soils we depend on to grow our food.

How can **weathering** and erosion be all that? The investigations you have done so far give some clues about these processes. Remember, in **physical weathering**, rock breaks down into smaller pieces. The smaller pieces are called sediment. Erosion transports sediment to a **basin** by water, wind, or ice.

The abrasion, or scraping, of blowing sand helped carve and smooth this sandstone canyon.

Physical Weathering and Round Rocks

Physical weathering occurs when large rocks break into smaller rocks of the same kind. When a rock, like granite, is broken, it may break into small pieces of the **minerals** that make it up, such as quartz and feldspar. But they are the same minerals that were in the original granite.

The sharp edges and corners of broken rock pieces wear away as they hit other rocks. This reshaping occurs naturally when rocks are hit by windblown sand or rock particles in moving water. The name for this type of physical weathering is **abrasion**. Abrasion also happens when falling rocks hit other rocks, breaking them apart.

When you observe beach sand or sand in a riverbed, you can see smooth, polished sand grains. Waves and flowing water rolled these sand grains around, causing them to hit each other. You observed sand particles in the stream table bouncing and hitting other grains of sand as they moved along. The water carries rocks that bump off the rough edges on other rocks. The farther the sand grains are carried by water, the smoother they get.

No water is involved when wind transports sediment. The sand grains bang into each other, creating a frosted, dull rounded surface. Beach sand, river sand, and dune sand are all similar in at least one way—the farther the weathered rocks travel and the more they get banged around, the smaller they become.

Ice Wedging and Rock Falls

When ice freezes, it expands with great force. You saw what happened when water in a jar froze and expanded. The force shattered the jar! Ice expansion naturally causes physical weathering when water gets into tiny cracks in a rock. At night, temperatures fall, and the water freezes, expands, and presses against the surrounding rock. The crack gets bigger. During the day as the temperature rises, the water thaws and seeps farther down into the crack. Night comes, and the water freezes again. With repeated freezing and thawing, the crack becomes larger. Eventually, pieces of the rock break off. Ice wedging can break rocks off the side of a cliff. These **rock falls** make piles of jagged rocks, called **talus**, at the base of cliffs. These rocks fell because a tiny crack kept growing until the piece broke off. You may have seen the results of ice wedging in your neighborhood. Ice damages concrete sidewalks and curbs, and wedges flakes from bricks.

Plant roots also cause weathering, by growing into cracks. The roots expand as the plant grows, breaking the rocks apart. You may have seen tree roots that lifted and broke a sidewalk or even cracked the foundation of a house.

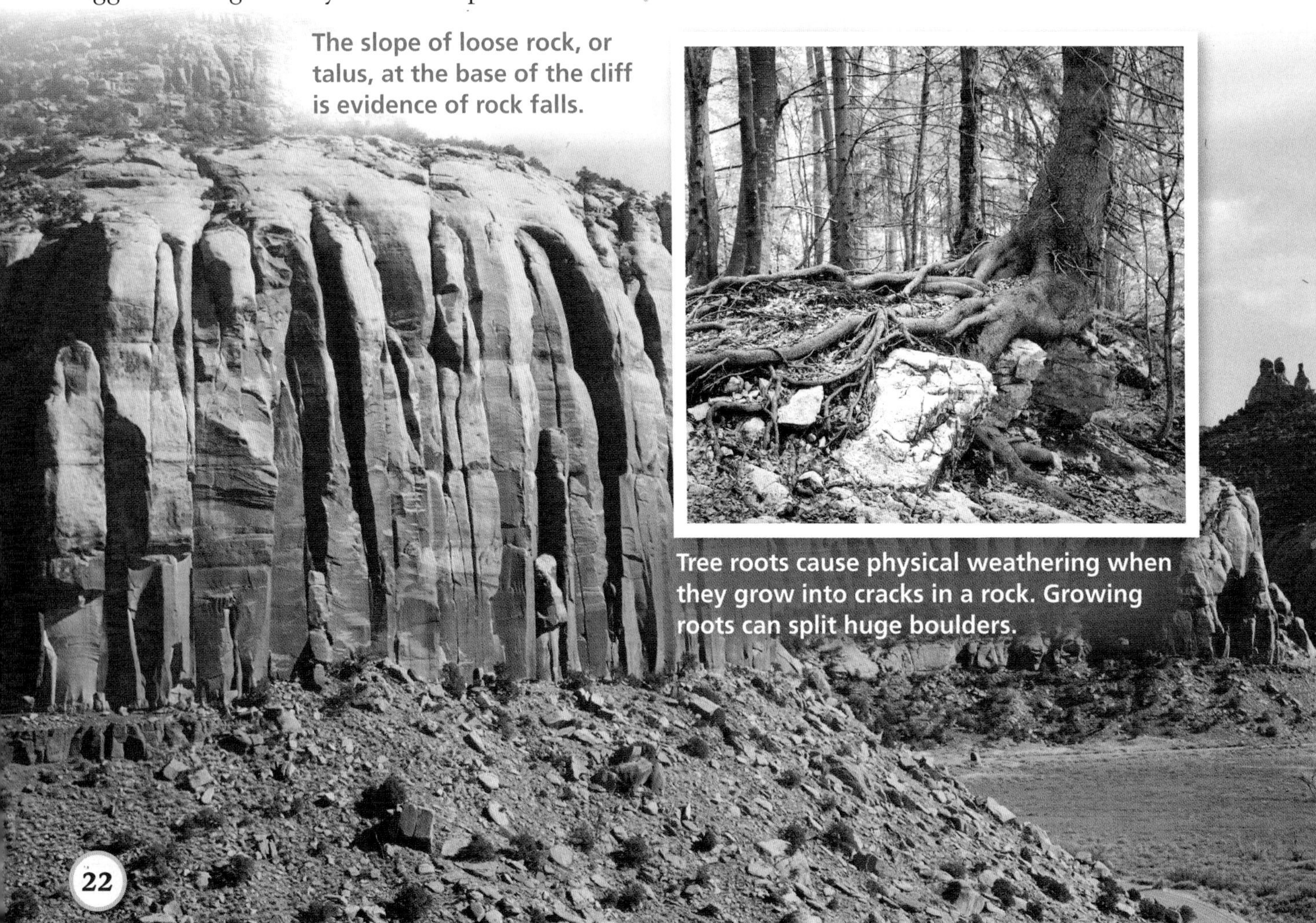

The slope of loose rock, or talus, at the base of the cliff is evidence of rock falls.

Tree roots cause physical weathering when they grow into cracks in a rock. Growing roots can split huge boulders.

Vast karst landscapes are found in the humid tropics of Southeast Asia. The towers, pinnacles, and cones are caused by chemical weathering.

Chemical Weathering

Physical weathering is not the only way to break down rocks. Remember how the limestone fizzed when you put a drop of acid on it? During that **chemical reaction**, acid dissolved a tiny amount of rock. This same process takes place naturally. Tiny amounts of carbon dioxide in the air dissolve in falling raindrops. The solution is a very weak acid called carbonic acid. This acid is too weak to make limestone fizz. But each slightly acidic raindrop dissolves a few **molecules** of limestone. Over thousands of years, limestone will slowly wear away because of **chemical weathering**.

Decaying plant material, lichens, and plant roots also produce carbon dioxide and other weak acids. These acids dissolve in rainwater as it moves through **soil** and into cracks in rock. Limestone caves such as Mammoth Cave in Kentucky formed over millions of years as chemical weathering dissolved limestone. This weathering creates landforms called **karst topography**. Sinkholes and caves are karst landforms and are found in many areas, including Kentucky and Florida. Dramatic limestone pinnacles found in Asia are also examples of karst topography. Because acids need moisture to form, karst topography is found in humid environments. In dry locations, limestone tends to form steep cliffs, like the Redwall Limestone in the Grand Canyon.

Weak acids also weather granite, though much more slowly than limestone. Granite is made of several common minerals, mainly quartz, feldspar, and hornblende. Quartz is very resistant to chemical weathering. Feldspar is easier to break down. Acid slowly weathers feldspar into clay particles. Without feldspar to hold the granite together, the quartz **crystals** fall out and become sand. The toughness of quartz is the reason so much sand is mostly quartz.

Differential Erosion

You saw **differential erosion** in action in the stream tables that had a layer of clay between two layers of sand. The water easily eroded away the top layer of sand. The clay layer resisted erosion. As long as the clay layer was solid, it protected the bottom layer of sand. This is called differential erosion, because the layers erode at different rates.

This large mushroom rock near Lees Ferry on the Colorado River in Arizona is the result of differential erosion.
Warning: Rock structures of this kind can be hazardous. Be careful around them.

Once hidden below Earth's surface, Devils Tower now looms above the Wyoming prairie. Erosion has worn away the soft sandstone and shale that once overlaid this plug of hard volcanic rock.

The Niagara River tumbles over a hard limestone layer to form a waterfall 50 m high. Millions of gallons of water plunge downward every minute.

Differential erosion happens any time soft rock is eroded away, leaving harder rock behind. Much of the scenery in the Grand Canyon is due to differential erosion. Devils Tower in Wyoming consists of hard rock that was once surrounded by softer rock. Over the past 1 to 2 million years, the softer rock weathered and eroded away. The column of hard rock still stands.

Niagara Falls on the New York–Canada border is another example of differential erosion. The water going over the falls erodes the edge of the thick, soft **shale** layer under a hard limestone layer. This undercuts the limestone, causing it to give way. The photo above of American Falls, a part of Niagara Falls, shows huge limestone boulders that have fallen. For the past 10,000 years, the falls have moved upstream, eroding the rock at the plunging edge of the fall, at an average rate of about 1 meter (m) a year.

Wind Erosion and Rain Forests

How could erosion in Africa help the Amazon rain forests, all the way across the Atlantic Ocean? Windstorms on the Sahara Desert carry dust high into the **atmosphere**. High-altitude winds carry the dust west across the Atlantic Ocean. Sometimes it reaches the rain forests of South and Central America. The soil in rain forests is normally poor in nutrients, and the dust from Africa provides many of the nutrients that the rain forest plants need to survive.

Dust storms can carry silt and other fine particles long distances, even across the ocean.

There is also a negative side to all this dust. Some of the dust settles on coral reefs in the sea. The dust carries bacteria and fungi that can kill or weaken the coral. When the dust settles over populated areas, it can trigger asthma and other respiratory diseases. Different places in the world are affected by different dust sources. Dust from China sometimes reaches people living in California!

Weathering and erosion created the Grand Canyon and many spectacular landforms around the world. The spires and **hoodoos** of Bryce Canyon in Utah, the rugged Badlands in South Dakota, and the rounded Blue Ridge Mountains extending from West Virginia to Pennsylvania are all products of weathering and erosion. Mammoth Cave in Kentucky, the world's longest cave system, was created by weathering and erosion. All these wonders were once solid rock.

Weathering and erosion produce sediments that can form new rocks, create soil, change landforms, and affect air quality. Think about that the next time you feel smooth, rounded sand between your toes at the beach, see a crack in a sidewalk, or eat a carrot!

Think Questions

1. **Choose one of the photos of rock formations in this article. Make a sketch of the formation. Label the layers that you think might be hard rock and those that might be softer. What is your evidence?**
2. **Describe the processes you think might have produced the mushroom-shaped rock in the photo.**
3. **Think about your community. Give at least one example of where you have seen these processes.**
   - **Weathering**
   - **Erosion**
   - **Differential erosion**

Soil Stories

Without soil, most of the food we eat would not exist. That includes fruits, vegetables, hamburgers, spaghetti, and even pizza! Soil is essential to most terrestrial life on Earth.

What Is Soil?

Soil is the loose top layer of Earth's **crust**. It is a mixture of mineral and organic materials, air, and water. One ingredient of soil is weathered rock. When solid rock is exposed to air, water, and living organisms, it breaks down into small pieces like sand, silt, and clay.

But soil is much more than just sand, silt, and clay. It's alive! One gram of rich soil (about the mass of a large paper clip) can contain 1 million fungal cells and 1 billion bacteria. In addition, soil contains insects, earthworms, and many other organisms. These organisms feed on dead plants and animals. They break them down to create an important ingredient in soil called **humus**. These organisms also loosen the soil, so that water and plant roots can move through it more easily. Weathered rock particles and humus are the main components of soil.

The composition of a soil sample can vary widely from one location to another.

Lichens growing on rocks produce weak acids that can weather and crumble the rock surface.

Soil Formation

All the organisms in the soil and humus give off weak acids. Rain is also slightly acidic, because raindrops dissolve carbon dioxide and pollutants from the air. These liquids soak down through the soil. They help break down underground rock through chemical weathering. This process releases phosphorus, calcium, and other chemicals from the rock. All these chemicals become nutrients for plants and the animals that eat them.

When plants and animals die, humus-forming organisms in the soil decompose them. The nutrients in the humus are then used by living plants. Plants need phosphorus to make DNA, so that cells can divide. They use calcium to make cell walls. Humus also provides nitrogen, an element needed to make proteins. Without nitrogen, plants cannot grow. Besides providing nutrients for plants, humus helps hold soil in place. It also helps hold water in the soil.

It can take more than 200 years for 1 centimeter (cm) of soil to form. Soil formation may start with a colony of lichen on bare rock. Lichens are organisms made up of fungus and algae living together. The lichen releases weak acids, which slowly break down the rock on which it lives. The weathered rock plus a tiny amount of humus from decaying lichen creates a speck of soil. Once a few specks accumulate, there is enough soil to support moss or other tiny plants, which also add humus to the soil. It can take thousands of years to produce enough soil to support a small shrub.

Often, human activities such as construction, farming, or logging expose the soil. Then soil erosion can quickly occur. It can take centuries to renew this precious resource, but soil can be carried away in a few hours during a heavy rain. Terracing, contour farming, and replanting grass or trees on bare soil slows the movement of water and dramatically reduces soil erosion. Plastic barriers around construction sites help slow soil erosion.

Types of Soil

Just like rocks, soils are not all the same. Analyzing soil gives you clues to how it formed and where its materials came from. Soil can help you tell the geologic story of a place. Here are some of the basic kinds of soil.

Sandy soil. Some soils have mostly sand-sized grains of minerals and rock. If you rubbed sandy soil between your fingers, it would feel gritty. Because of large spaces among the grains, sandy soil has good drainage, meaning that water can flow easily through it. But water moving too quickly carries away most of the nutrients. This makes sandy soil poor for growing most plants.

Clay soil. Soils with lots of clay have plenty of mineral nutrients. But because there is very little space between the clay particles, roots cannot grow well in clay soil. Water cannot move through clay soil easily, so it has poor drainage. Some plants will drown if clay soil becomes saturated with water. Clay soil is not very good for growing most plants. Wet clay soil feels slick when you rub it between your fingers.

Silty soil. One of the best soils for growing plants has lots of silt. Silty soil contains rich nutrients, has fairly good drainage, and usually has plenty of humus. If you rub a bit of silty soil between your fingers, it would feel smooth and barely gritty.

Loam soil. Many soils are made up of several sizes of soil particles. The amounts of clay, silt, and sand vary from one loam soil to another. Because loam soil contains sand and silt, water can move through it easily. Because loam soil has good drainage and plenty of humus and mineral nutrients, it is good soil for gardening or farming.

While almost any plant will grow well in a loam soil, some plants require well-drained soil (sandy loam or loam). Others thrive in clay soil with a lot of loam. Pine trees, for example, require well-drained soil and will not survive well in clay soils.

Combination soil. Most soils contain several different particle sizes. **Pedologists** (soil scientists) describe soils in terms of how much of each size of particle the soil contains. For example, if a soil contains a lot of silt, some sand, and very little clay, it will be called a sandy silt soil. If the soil is mostly silt, but also has quite a bit of clay and sand, it is called a silty loam soil.

Soil erosion can be caused by human activity like farming. Letting animals overgraze, or feed in an area for too long, destroys the plants that hold the soil in place.

Regional Factors

The type of soil that forms in an area depends on many factors. Here are five important factors.

Parent material. Soil forms from a parent rock or sediment. For example, granite is a parent material with lots of quartz crystals. It will weather to form sandy soil. Sandstone will also weather to form sand. Shale and mudstone weather to form clay or silt.

Climate. Temperature, moisture, wind, and sunlight all contribute to differences in soil. Warm temperatures and moisture encourage plant growth and chemical processes that break down rock fairly quickly.

Topography. Soil formed in hilly or mountainous areas is easily eroded into valleys, plains, or other basins by water, wind, and **glaciers**. The deep, rich soils of the Great Plains include deposits of sediments from the Rocky Mountains. Many soils of the Midwest were carried and deposited by glaciers within the past 20,000 years. The soils of the Central Valley in California, Sun Valley in Arizona, and the coastal plains of the East Coast, as well as other valleys, were mostly eroded from the surrounding mountains and hills.

Some soils are better able to provide the water, nutrients, and physical conditions that plants need to thrive. These trees growing in clay soil could not survive a dry period.

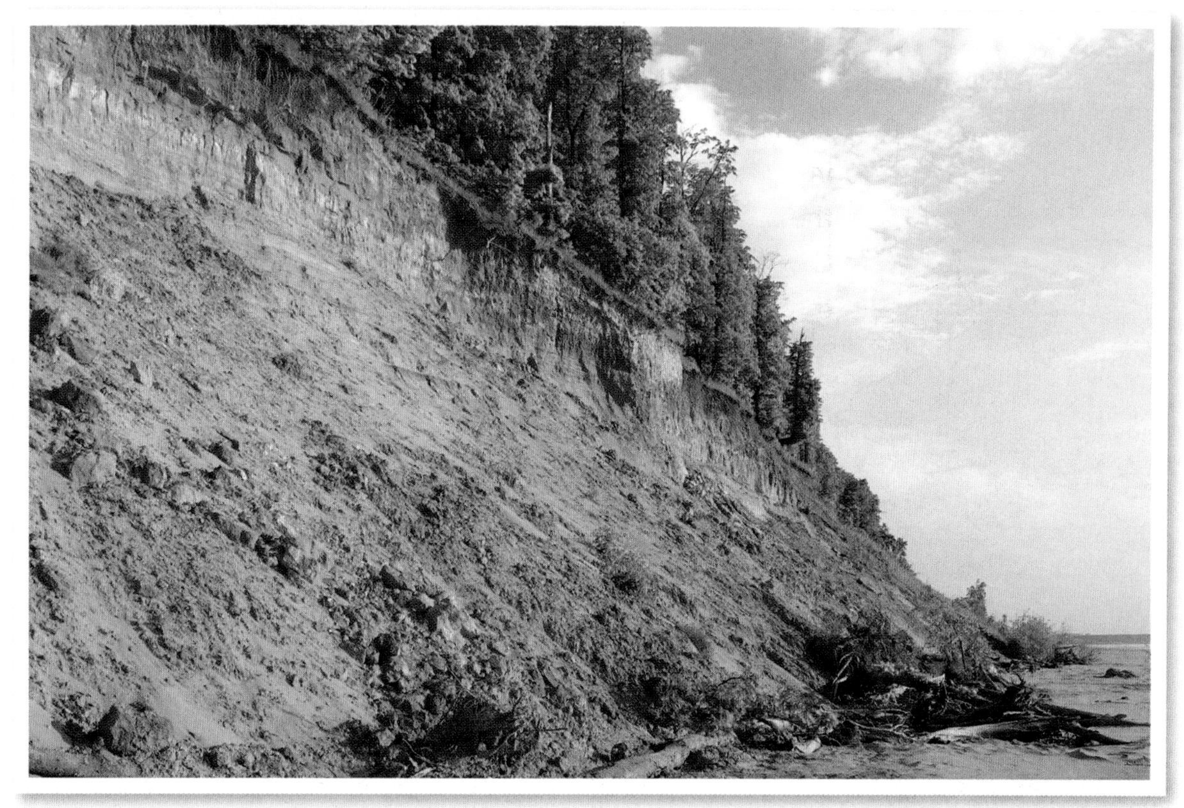

Soil along a coastline is particularly vulnerable to weathering and erosion by water. The steeper the land and the faster the water flow, the greater the erosion.

As you saw in the stream table, water sorts soil particles. The river bottom and areas where the river flowed in the past are usually very sandy, because the river carried the clay and silt particles away. The soils in the areas near the river, where the water spread out and deposited the smaller particles, contain more silt and clay.

Biological factors. Small animals burrow into and loosen the soil, allowing water and plant roots to penetrate more easily. Plant roots can enlarge cracks in rocks. Plants, microbes, and other organic matter can produce weak acids that cause rocks to break down to form soil.

Time. The amount of time it takes for all the factors to interact affects the type of soil that forms.

Did You Know?

Fungi and bacteria in the soil are constantly competing for organic matter. They use it as food. Some fungi produce chemicals called antibiotics that kill bacteria. It is microscopic chemical warfare! Many of the antibiotics, such as penicillin, that we use today were discovered in soil fungi.

The fuzzy mold on this orange consists of fungus that may also be found in soil.

Soil Profile

Digging into the soil reveals distinct layers of soil called horizons.

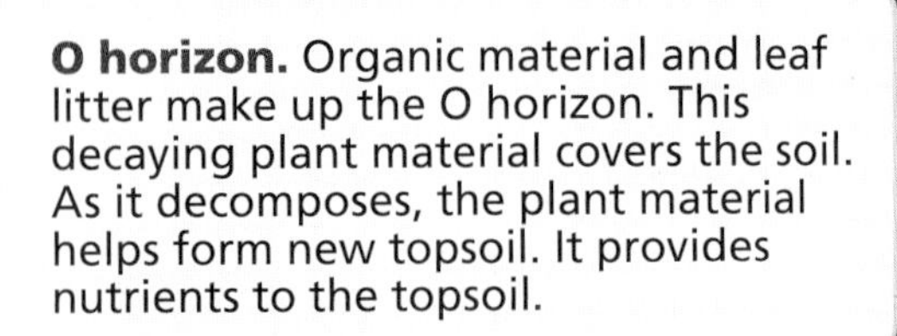

O horizon. Organic material and leaf litter make up the O horizon. This decaying plant material covers the soil. As it decomposes, the plant material helps form new topsoil. It provides nutrients to the topsoil.

A horizon (also called topsoil). Topsoil is the top horizon of soil in a profile. The humus in this layer usually colors the A horizon dark brown. Most humus, nutrients, roots, and organisms necessary for healthy soil are found in this layer. If a lot of the topsoil erodes away, plants will not grow well in what is left.

B horizon. Subsoil (B) is usually a lighter color than the A horizon. The subsoil holds mostly sand, silt, and clay. It has almost no humus. Usually only the roots of large plants penetrate to the B horizon. The chemical changes that take place here create nutrients that can, over time, become available to plants. Eventually, the top of the B horizon can change enough to become part of the topsoil.

C horizon. Soil develops from a parent material. The parent material can be sediments that were deposited by wind, water, or glaciers. It could be the weathered bedrock that became the A and B horizons. Below the C horizon with its parent material is solid bedrock.

Soil Horizons

When you dig a deep hole or find a place where a road or stream has cut through the layers of soil, you can see a **cross section**. This view is called a **soil profile**. A soil profile usually has a number of layers called **horizons**. Horizons are somewhat like the rock layers you see in the Grand Canyon, but they are made of soil, not solid rock. Soil horizons usually differ in color, texture, and chemical and mineral content.

Most horizons are made of weathered rock material. Soil scientists and geologists use capital letters to identify types of horizons.

Can you see the different soil layers, or horizons, in this cutaway view? Look for different colors and different-sized particles.

You may find that the soils where you live have really interesting stories to tell, once you know how to look for clues. Does your soil contain volcanic ash? Does the soil type show where ancient rivers might have been? Was the soil deposited by glaciers?

FOSS **If you would like to find out more about soil, especially the soils where you live, check out the resources on FOSSweb.**

Where in the World Is Calcium Carbonate?

Millions of caves lay hidden under mountains, canyons, plains, and water. These geologic features have developed over thousands of years.

The deepest, longest, and most spectacular caves are carved in limestone. What kinds of interesting stories do you think they can tell?

What Is Limestone?

Sedimentary rocks cover about 75 percent of Earth's surface. Limestone makes up about one-quarter of these rocks. The major component of limestone is calcium carbonate ($CaCO_3$). Calcium carbonate is the chemical name for the mineral **calcite**. Calcium carbonate is also the main mineral in the shells of many invertebrate organisms. It is in the hard structure of coral and clam shell. It is the main ingredient in marble, chalk, and tufa.

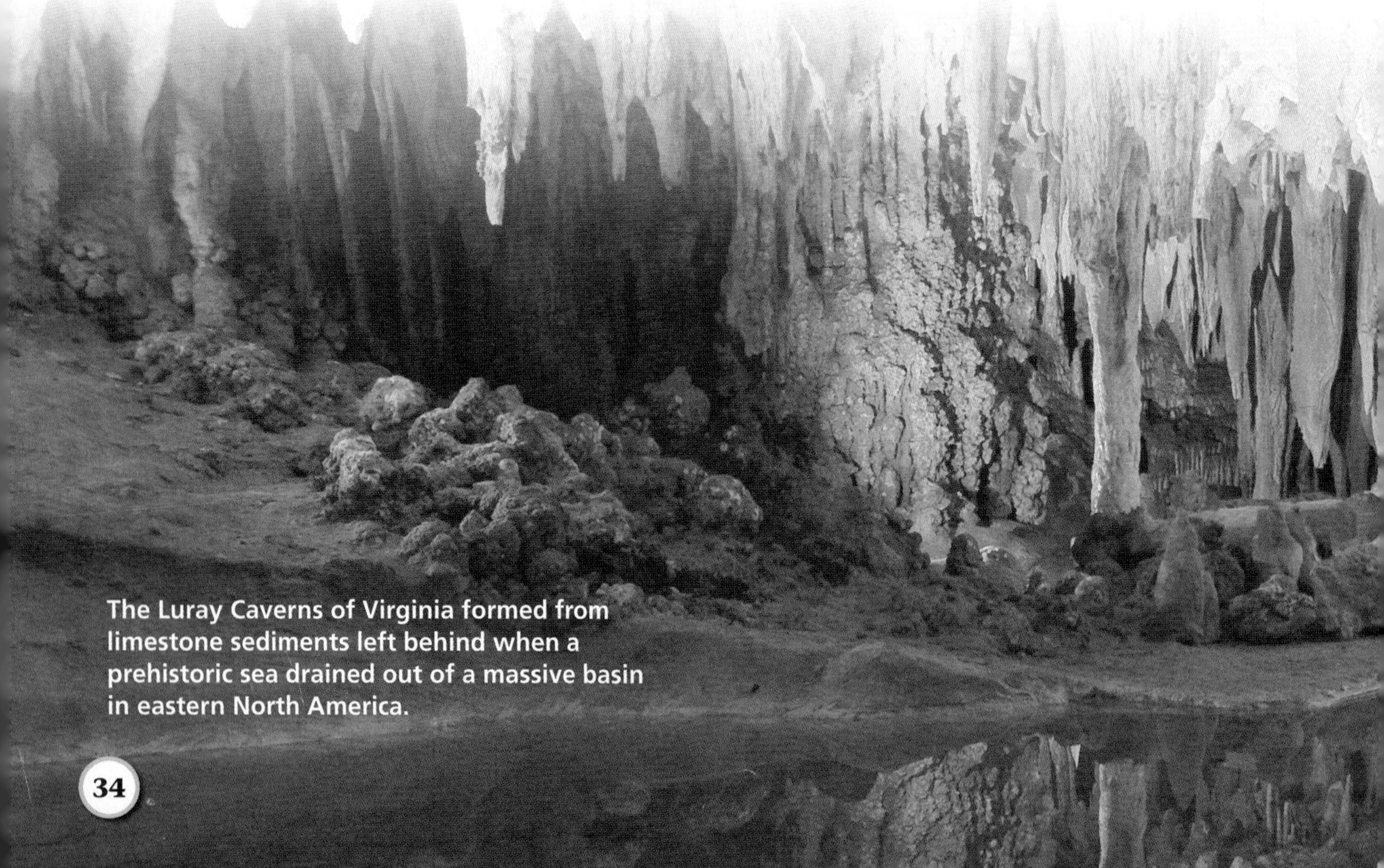

The Luray Caverns of Virginia formed from limestone sediments left behind when a prehistoric sea drained out of a massive basin in eastern North America.

Calcite is composed of three elements, calcium, carbon, and oxygen. In its purest form, it is colorless. Trace elements can make it white, pink, red, yellow, green, blue, pale violet, black, or brown. Some calcite glows in ultraviolet light. Iceland spar is a transparent form of calcite that has an interesting property. If you look at something through Iceland spar, you will see a double image.

Calcite is a relatively soft mineral. It can be scratched by a knife, but not by a fingernail. Physical weathering easily breaks it into smaller pieces.

Calcite and other **carbonate** minerals are easy to identify. They all dissolve in acid. Calcite fizzes and releases carbon dioxide in weak hydrochloric acid. Most geologists carry a small bottle of hydrochloric acid to help identify calcite.

Calcite and limestone dissolve in areas with acidic **groundwater**. Huge caves can form where acidic groundwater seeps through limestone rocks. But if weathering quickly breaks down limestone, why is it so common on Earth's surface?

Pure, clear Iceland spar refracts light into two separate images. That is why it looks like you are seeing double.

Making Calcium Carbonate

Calcium carbonate results from chemical reactions in sea water. These reactions involve calcium and bicarbonate. Calcium in sea water can come from the chemical weathering of rocks such as granite. Bicarbonate can form when carbon dioxide in the air combines with rainwater. If the temperature is right, calcium and bicarbonate can react to form calcium carbonate.

Plants and animals can also create calcium carbonate. They take in calcium and carbonate from sea water. Then, they deposit calcite in their shells or release calcium carbonate into the water. These organisms include one-celled animals, oysters and clams, algae, and aquatic grasses.

Making Limestone

Eventually, the calcium carbonate and shells that these organisms produce collect on the ocean floor. They produce **ooze** that solidifies over time. It becomes limestone. Most limestone throughout Earth's history formed in relatively shallow, warm, tropical seas. These conditions still exist around tropical islands. There, calcium carbonate is easily available to marine animals and plants.

In cold water, calcium carbonate shells dissolve when the animals die. So nothing is left to form carbonate sediments.

The hard shells of oysters and other shellfish are composed mostly of calcium carbonate.

Varieties of Limestone

The one ingredient shared by all limestone is calcium carbonate. Geologists have described a variety of limestone. They often name the rocks for where they form or for the **fossils** and other materials in them.

Petroleum limestone is mainly composed of ooze and the microscopic shells of animals. This limestone forms in deep ocean water. The microscopic animals sink to the bottom and settle in the ooze there. Because deep water holds little oxygen, the organic material does not decompose. Over time, it changes to petroleum. These deposits can be hundreds of meters thick. It takes a lot of microscopic shells to build these deposits!

Reef limestone is mainly made of skeletons of marine organisms such as corals. Today, coral reefs grow in warm seas, around places such as the Bahamas and the Hawaiian Islands.

Coral reefs can form huge limestone deposits. The limestone contains fossils of all the reef organisms.

Much of the ocean floor is covered with a muddy form of calcite called calcareous ooze that over time will become limestone.

The beautiful blue-green sea water surrounding the Bahamas gets its color from bacteria that release calcium carbonate.

The famous White Cliffs of Dover, made of soft, fine-grained chalk, rise nearly 100 m and stretch along 16 km of England's south coast.

The Redwall Limestone in the Grand Canyon includes fossil corals and other reef-dwelling organisms. The fossils suggest that the Redwall Limestone was deposited in a warm, tropical sea.

Chalk is a powdery, fine-grained rock. It is almost pure calcite. It may be composed of the remains of algae called coccoliths. Coccoliths once lived in shallow seas. Chalk can also form from broken shells or precipitate from sea water. Much of the chalk on Earth was deposited on continental shelves around 70 million years ago. The white cliffs of southern England and the Austin Chalk that covers much of Texas formed then.

Coquina is made almost entirely of fossil bits. This limestone is mostly small shells and shell fragments, cemented together. The source material for coquina is similar to the sand on some beaches in the Florida Keys.

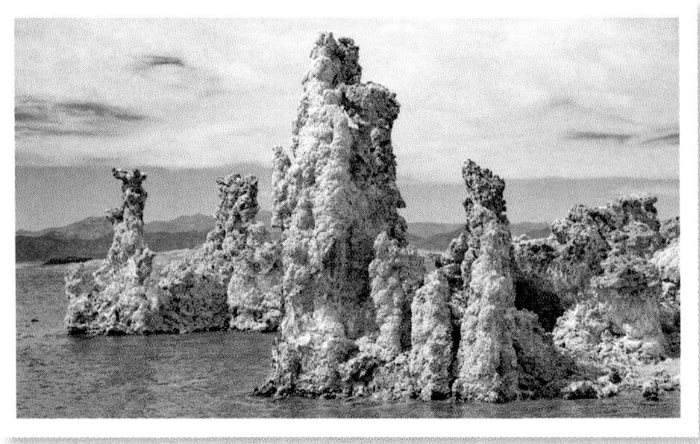
Tufa towers provide a unique environment for local organisms.

Tufa forms in cool carbonate-rich lake water. When calcium-rich spring water enters the lake, carbon dioxide is released and calcium carbonate precipitates.

Travertine is a calcium carbonate rock that is hard and dense. It is often deposited around **hot springs**, like those in Yellowstone National Park. In the Grand Canyon, travertine forms terraces in hot mineral springs.

Caves

When you saw a rock fizz under drops of acid, you witnessed chemical weathering. If you dropped the rock into a bucket of acid, it would slowly dissolve. Most water is slightly acidic. When acidic water runs through cracks in limestone for thousands of years, it eventually dissolves the limestone. Over time, acid weathering can form large underground caves.

Groundwater that contains dissolved limestone forms a carbonate solution. When this solution drips from cracks in the tops of caves, it reacts with air. Calcium carbonate can **precipitate**. The calcium carbonate crystals form spikes on the ceiling called **stalactites** and pillars on the floor of the cave called **stalagmites**. When stalactites and stalagmites meet, they form structures called columns.

Think Questions

1. **What are some properties of the mineral calcite?**
2. **What is a test that will reveal the presence of calcite?**
3. **Describe some rock formations that are likely to be made of calcium carbonate.**

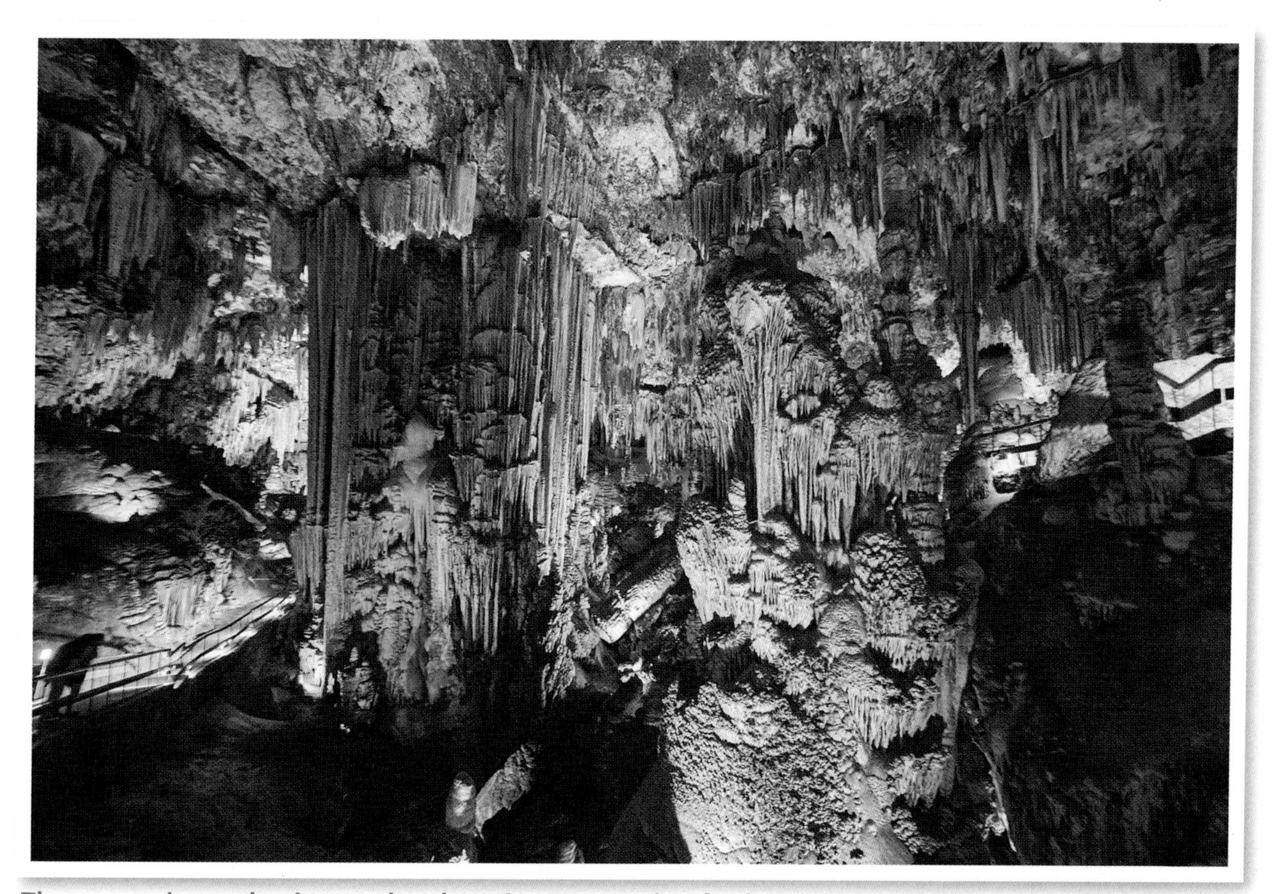

These massive stalactites and stalagmites are made of calcium carbonate that was dissolved in groundwater dripping into the cave. Trace elements and impurities produce color variations.

Water on Mars?

What do you think might have caused the Mars landscape below?

You have been looking at evidence from the environment to explain the story of each place on Earth. When you look at this image of Mars, what do you see?

Geologists have not been to Mars. But they can apply Earth landforming principles to think about the **formation** of landforms beyond Earth.

Uniformitarianism

Geologists use their observations about processes happening today, such as erosion and deposition, to infer events of the past. The idea is that the processes we observe happening today are the same processes that have always been at work. The processes are uniform. **Uniformitarianism** means that we can use the present as the key to the past.

So what about those landforms on Mars? Uniformitarianism says that we can apply processes on Earth to the past. The rule applies even if those rocks are in a very distant but similar environment, like Mars.

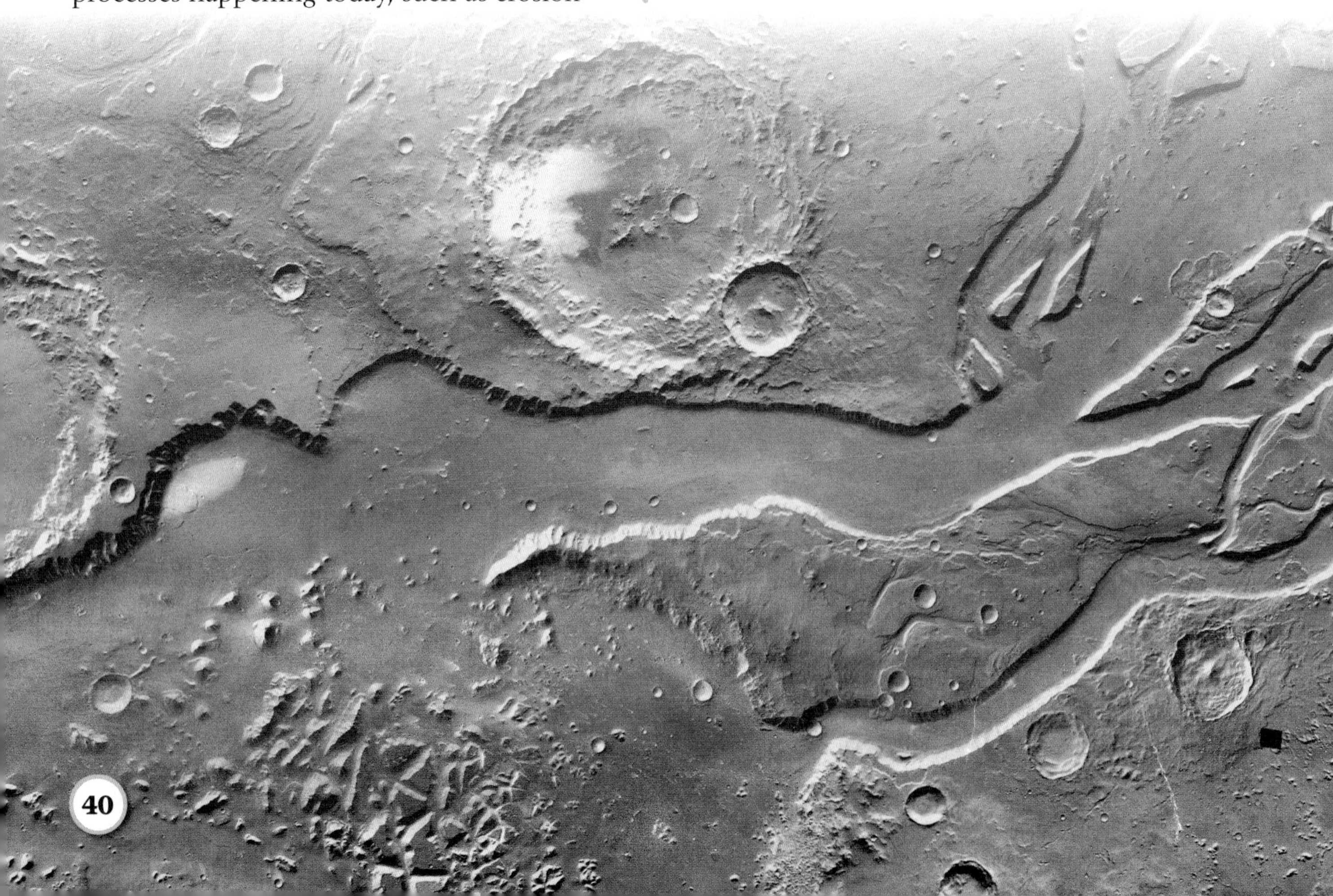

Early Evidence

Early astronomers and geologists were already thinking about what shaped Mars. In 1877, Giovanni Schiaparelli (1835–1910), director of an Italian observatory, observed Mars through a telescope. He made detailed drawings of his observations. He recorded lines that he called canals.

Astronomers were excited about the possible existence of water on Mars. Water is considered essential for life. So in looking for evidence of life beyond Earth, liquid water is an exciting find. However, most astronomers were skeptical. Very few could confirm the presence of canals.

Evidence from Spacecraft

In 1969, two Mariner spacecraft flew by Mars. Finally, we got a close look at the surface. But the images they collected did not answer the question about water on Mars. In 1971, *Mariner 9* launched. It spent 349 days in Mars orbit, getting 7,329 images that cover about 80 percent of the planet. It observed ancient riverbeds, craters, extinct **volcanoes**, canyons, polar deposits, evidence of wind-driven deposition and erosion, weather fronts, ice clouds, dust storms, and fogs. Some landforms appeared to be shaped by water flow. They suggested that liquid water once flowed on the surface of Mars. The evidence was inconclusive, however.

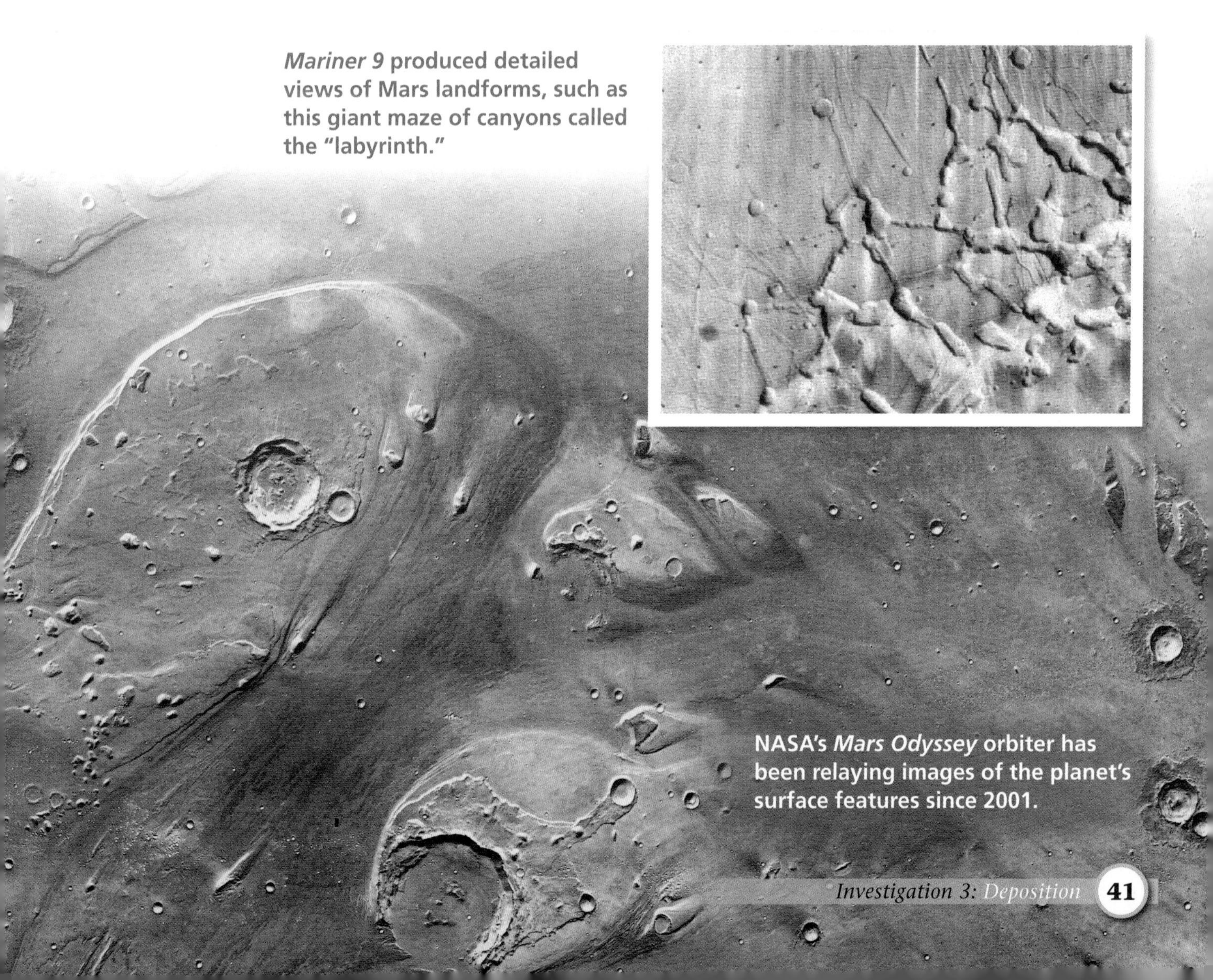

Mariner 9 **produced detailed views of Mars landforms, such as this giant maze of canyons called the "labyrinth."**

NASA's *Mars Odyssey* orbiter has been relaying images of the planet's surface features since 2001.

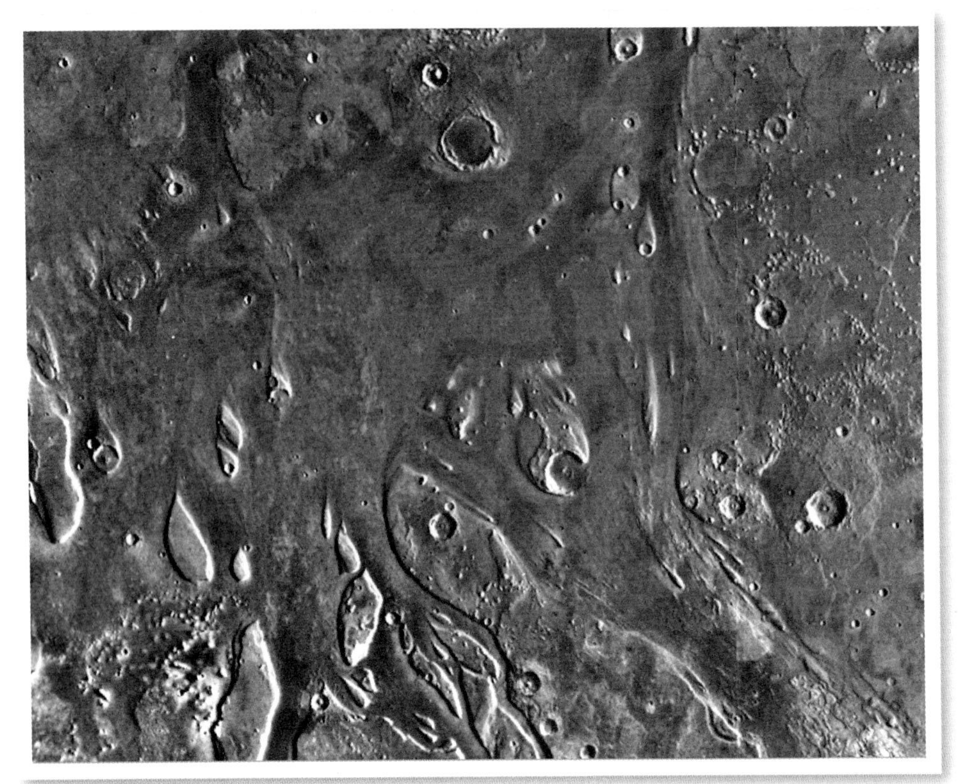

Scientists think *Pathfinder's* landing site, one of the rockiest parts of Mars, may once have been a floodplain. Do you observe evidence that water might have flowed here?

Evidence from Landers

In 1997, the *Mars Pathfinder* landed on Mars. Scientists studied thousands of images from the lander. They reported that liquid water was probably once present on Mars. Their evidence comes from images of the surface. These images seem to confirm that a giant flood left boulders, cobbles, and pebbles throughout the Pathfinder landing site.

Scientists also found evidence for a mineral known as maghemite. Maghemite forms in water-rich environments on Earth. Scientists suggest that maghemite could form in the same way on Mars.

The Pathfinder team discovered reddish rocks with evidence of oxidation on their surfaces. Oxidation is a type of chemical weathering that occurs on Earth. It causes iron to rust and turn reddish in the presence of water. So scientists inferred that water must have existed on Mars's surface at one time.

With all this evidence of water, new questions arose. Where did the water go? Perhaps it all evaporated. Perhaps some soaked into the ground or was frozen at the Martian poles.

In 2006, an analysis of Mars Global Surveyor data suggested that liquid water still flows on Mars in short events. This raises more questions. How much water could be below the surface? Is there a moist environment underground that could support life?

Confirmed Water

In 2008, a Mars lander dug into the planet's surface. It hit frozen soil about 5 centimeters (cm) deep. The arm scooped out soil samples, heated them, and identified water vapor from the sample. This was the first confirmation of water on another planet.

But not all water environments can support life. After looking for years, NASA confirmed deposits of carbonate on Mars in June 2010. Why is carbonate important? As you learned in class, carbonate forms in water and reacts with acid. Finding a deposit of carbonate means that Mars water was not acidic. It might have supported life in the past.

Did You Know?

As of 2017, there are five active NASA rovers or surveyors exploring Mars: *Mars Reconnaissance Orbiter, Opportunity*, *Mars Odyssey*, *Curiosity*, and *MAVEN*.

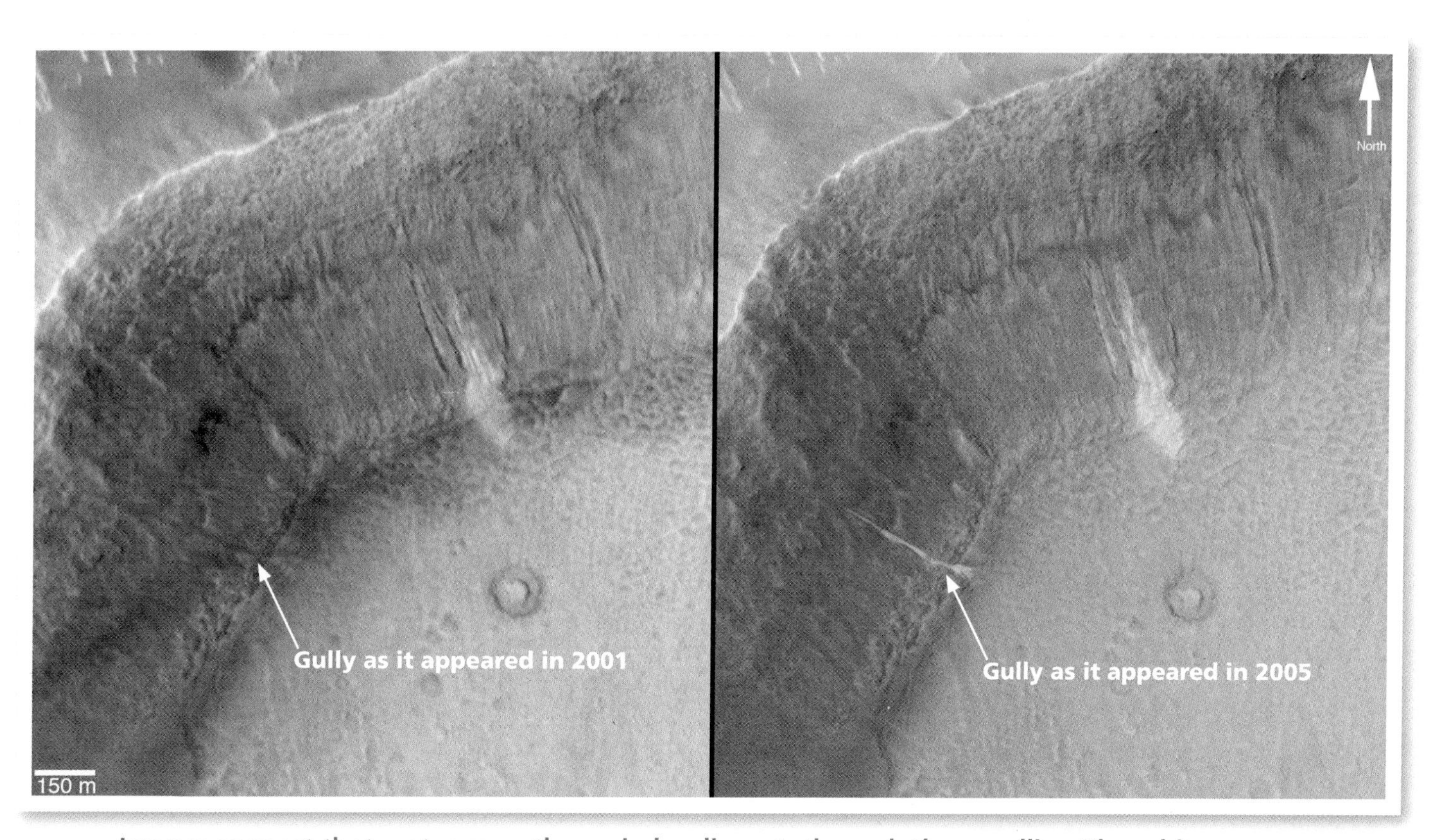

Images suggest that water recently carried sediments through these gullies. The white area appears to be sediments deposited by water.

This false-color image was created by a computer model that compiled data from several Mars missions. The dark streaks shown here, at Hale Crater, are roughly the length of a football field.

In 2012, the rover *Curiosity* landed on Mars to collect more evidence about Mars' past. And in 2015, the Mars Reconnaissance Orbiter analyzed areas on mountains where streaks had been observed. The streaks seemed to increase and flow during warm seasons. The chemical analysis revealed salts that only form in water, confirming that the streaks are formed by water in the warm seasons.

Where is the water coming from? Is it underground or does it condense from the atmosphere? Scientists still aren't sure. As you can see, every discovery on Mars leads to more questions.

People will eventually visit other parts of the solar system. Until then, scientists will continue to map the surface of the planets using probes and telescopes. They will send probes to collect and test surface material. The new data will provide better knowledge of the ancient history and current processes at work on solar system objects.

For more information about Mars exploration, visit FOSSweb for a link to the NASA website.

Think Questions

1. **After reading this article, what kinds of rocks would you expect to see on Mars? What evidence do you have?**
2. **What clues can you find in this article to explain why Mars is called the red planet?**
3. **What ideas and observations about Earth have scientists used to support the evidence that water might have existed on the surface of Mars?**

A Fossil Primer

Look at the landscape around you. You might ask, How did these rocks get here? Has it always looked like this? What's the story of this place?

If you know how and where to look, you can find the answers to questions about Earth's history written in the rocks. Geology provides the story of Earth's 4.6-billion-year history.

While exploring the Grand Canyon, John Wesley Powell (1834–1902) recognized that the rocks could tell the canyon's story. Sedimentary rocks, like those in the Grand Canyon, give us many clues to the story of a place.

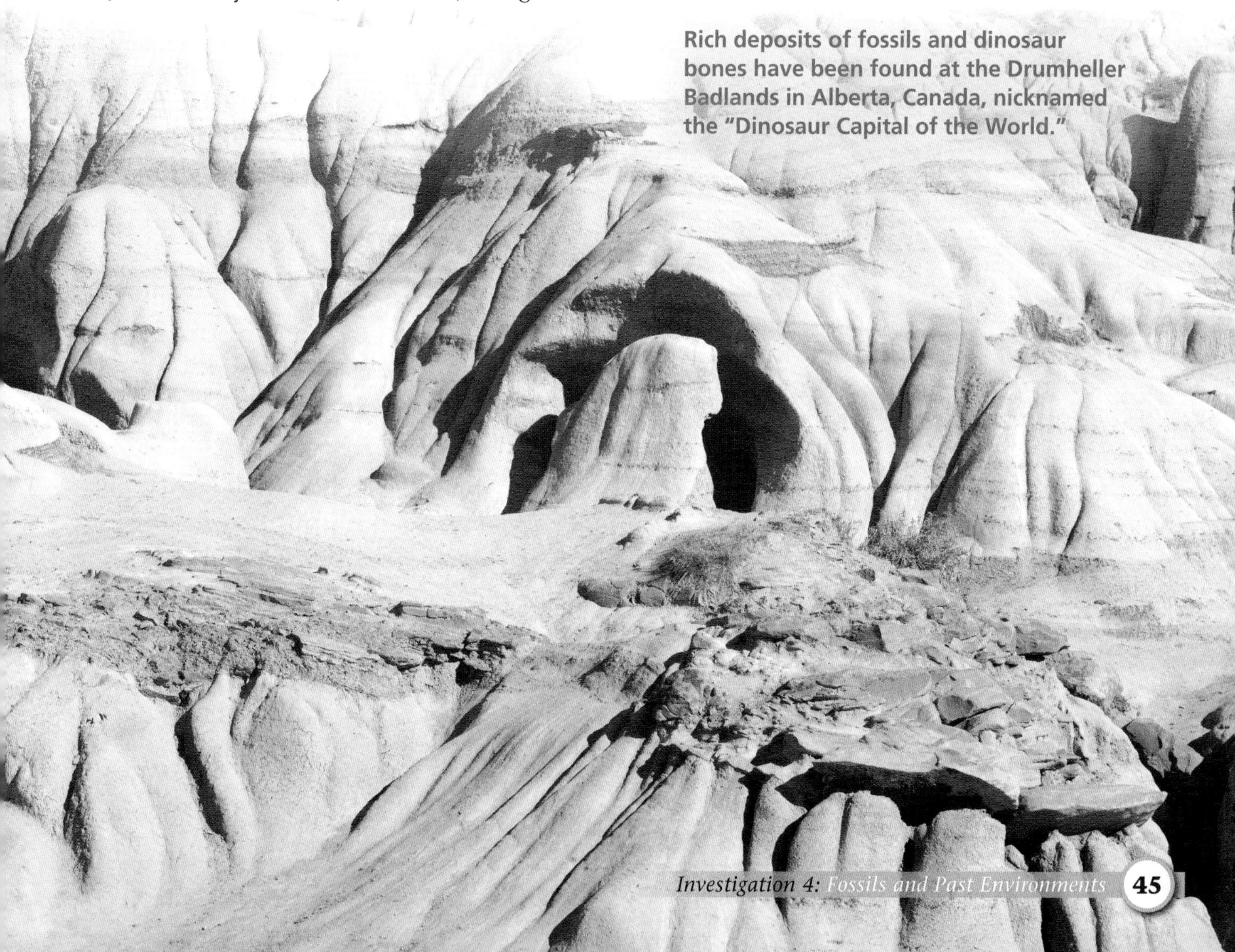

Rich deposits of fossils and dinosaur bones have been found at the Drumheller Badlands in Alberta, Canada, nicknamed the "Dinosaur Capital of the World."

Preserved animal tracks are trace fossils that provide evidence about the size of an organism and how it walked. What animal might have made this imprint?

Information from Sedimentary Rocks

- Sediments from ancient rocks, weathered, eroded, and deposited here.
- Layered rocks formed from sediments that were deposited flat and horizontal (the **principle of original horizontality**).
- Rocks in lower layers are older than rocks above them (the **principle of superposition**).
- Processes that form rocks and landforms today are the same as those that formed rocks and landforms in the past (uniformitarianism).
- Types of rock can tell us about the environment in which they formed.
  - Limestone forms from sediments deposited in water.
  - Sandstone forms from sediments deposited in sandy areas, like beaches, deserts, and dunes.
  - Shale forms from sediments deposited in calm, muddy environments, like swamps.
  - Some sedimentary rocks contain fossils. Fossils add details to the story of a rock's environment.

Evidence of Ancient Life

Fossils are found in rocks and are evidence of ancient life. The two largest groups of fossils are body fossils and trace fossils.

Body fossils preserve parts of the organism. Most commonly, the hard parts of the organism, such as shells, bones, and teeth, become fossils. Sometimes the soft parts are found, but they are rare. The conditions have to be just right to preserve the soft parts.

Trace fossils preserve evidence of the activities of organisms. These are most often footprints and tracks, tail-drag marks, burrows, impressions, and droppings.

How Do Fossils Form?

When you pick up a fossil, what are you looking at? It might look like a snail, but it also looks like a rock. There is no soft snail body inside.

When an organism such as a snail dies in normal conditions, the soft body parts decompose through the action of scavengers, bacteria, and oxygen. The hard shell breaks down through weathering, both chemical and physical. For most organisms, all evidence of their existence is lost. But on rare occasions, the snail ends up in an environment that preserves it.

Worthenia tabulata probably lived in a shallow, tropical sea. When this individual snail died, its shell sank to the bottom. For some reason, it was soon buried by sediments that isolated it from bacteria and oxygen. The sediments also protected the shell from being broken into small pieces or dissolved.

Over the next few million years, several things could have happened to the snail. Mineral-rich water might have slowly replaced the calcium carbonate in the shell with a harder mineral. The sediments around the shell might have hardened. Acidic water could later have dissolved the shell, leaving a shell-shaped space in the hardened sediments. The shell-shaped space might have filled with new, hard minerals from the surrounding water.

A snail fossil might lay buried for millions of years. Eventually, the surrounding rock layer reaches the surface. As the surrounding rock weathers and erodes, the fossil of the snail becomes exposed. If it is harder than the surrounding rocks, it might tumble out, just waiting for a geologist or rock collector to find it.

This fossilized snail, *Worthenia tabulata*, lived during the Pennsylvanian period, about 300 million years ago.

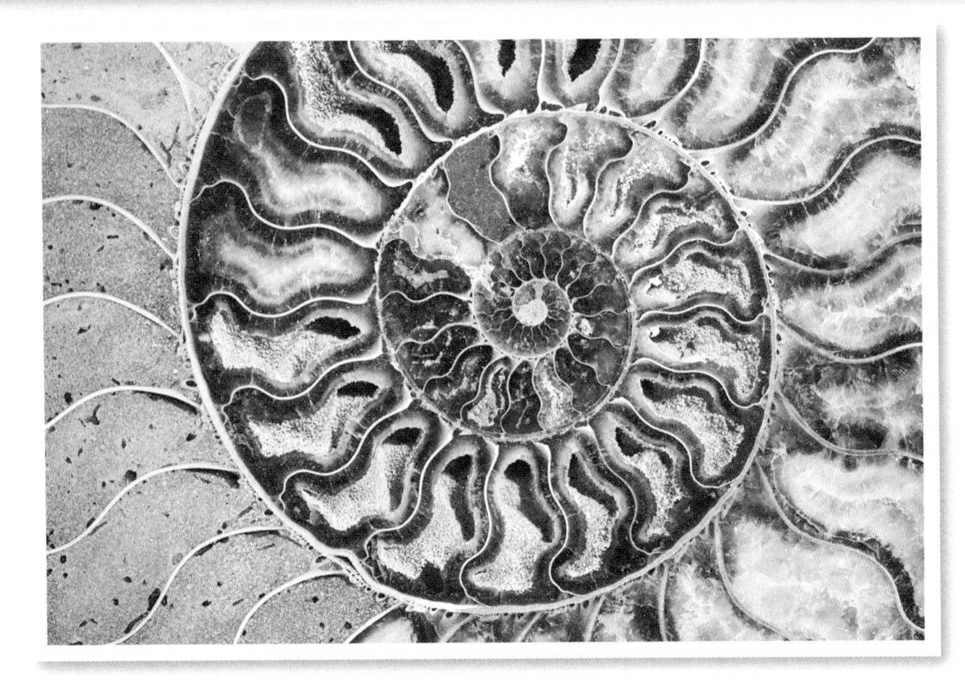

Slicing open this spiral ammonite fossil reveals that the inner chambers of the shell structure are lined with crystallized mineral deposits.

Trace fossils like footprints are even more rarely preserved. Footprints are left on exposed surfaces. So wind and water usually quickly erase them. Occasionally, they are covered and buried by soft sediments before they are disturbed. When the sediments are buried and turn to rock, the imprint is preserved. Millions of years later, the trace fossil will be exposed as the layers around it erode away. Because trace fossils and the surrounding rock are the same, these fossils are very fragile and are often lost as fast as the rock erodes.

Using Fossils to Read the Rocks

For centuries, people have been fascinated and puzzled by fossils. How they got into the rocks and, more importantly, *when* they got there raised many questions among scientists.

A French naturalist named Jean-Baptiste-Pierre-Antoine de Monet, Chevalier de Lamarck (Lamarck for short; 1744–1829), was a professor of zoology in the late eighteenth century. Lamarck used the principle of uniformitarianism, "the present is the key to the past," to explain what fossils are and how they form. He observed that, when present-day shellfish die, they are buried in mud and other sediments accumulated over long amounts of time. He realized that over a long enough time, the mud could turn to rock, and the shells stuck inside could become fossilized. Lamarck noticed that fossils of palm trees and corals found in a shale deposit near London were similar to some of the palms and corals that live in tropical environments today. Using uniformitarianism, he inferred that fossil corals lived in a prehistoric tropical environment similar to the environment where corals live today, even though the conditions around London where the fossils were found are nothing like a tropical environment today.

Lamarck contributed so much to the understanding of fossils that he is called the father of **paleontology**, the branch of science devoted to the study of fossils.

Every Fossil Has a Story

Even a small fossil has a story to tell. The type of fossil and the rock that surrounds it tell about the environment the animal lived in. The way it fossilized, whether replaced by minerals or filled in by sediments, can give clues to how the surrounding rock formed. When we compare the layers above and below, the principle of superposition tells us the relative age of the rock. When we consider the angle of the layers, the principle of original horizontality tells us what happened to the rock layers after the fossil formed.

Take Note

Pick one of the fossils you observed in class. Write a story that could explain its formation.

Think Questions

1. **Many dinosaur bones have been found. Why do we have little information about other parts of dinosaurs' bodies?**
2. **Does a fossil form before the rock it is found in?**
3. **What are some reasons that an organism might not turn into a fossil?**

Every fossil, from a complete dinosaur skeleton to a tiny snail, contains important information about how Earth's organisms and environments have changed over time.

Geologists can infer the order in which these rock layers formed. The oldest rock layers are generally underneath younger layers.

Rocks, Fossils, and Time

Earth is constantly changing—nothing on its surface is truly permanent. Rocks that are now on top of a mountain may once have been at the bottom of the sea.

To really understand the planet we live on, we must include the dimension of time.

We keep track of time with an invention most of us take for granted, the calendar. The calendar is based on the movements of Earth in space. One spin of Earth on its axis is a day. One trip around the Sun is a year. The modern calendar is a great achievement, refined over thousands of years.

People who study Earth's history also use a type of calendar, called the **geologic time** scale. It looks more like a book than a wall calendar. Rocks are the pages of this book. Some of the pages are torn or missing, and the pages are not numbered. Even so, geology gives us tools to help us read this book.

Rocks and Layers

We look into Earth's history by studying the record of events preserved in the rocks. Most rocks on Earth's surface are sedimentary. They formed from sediments, bits of older rock that were broken apart by water or wind. Some of these rocks contain fossils.

Sedimentary rocks form particle by particle and layer by layer. Gravel, sand, and mud settle on the bottoms of rivers, lakes, and the ocean. The layers pile up, one on top of the other. You saw this in the sedimentary basin you created in class. Superposition provides an important clue to the relative ages of rock layers and the fossils in them. A layer must be older than any layer on top of it.

Horizontal layers form when sediments settle from water or air (the principle of original horizontality). But many layered rocks are no longer horizontal. Why is that? The rocks moved from their original horizontal position. Events such as mountain formation and **earthquakes** can **fold**, bend, and tilt sedimentary rock layers.

Rock layers are also called strata. **Stratigraphy** is the science of layered rocks, which includes how layers relate in time. Figuring out the order in which a group of rocks formed, with the oldest on the bottom, creates a **relative time scale**.

Putting Events in Order

Think about these five historical events.

- TV becoming common in homes
- World War I
- World War II
- The Wright brothers' first airplane flight
- Astronauts landing on the Moon

Let's try to put these events in order using logic, common sense, and a little knowledge. We can infer that World War I occurred before World War II. We know (or were told) that millions of people watched Neil Armstrong on TV when he walked on the Moon. TV had not been invented when the Wright brothers flew their airplane. So we can order these three events: the Wright brothers' flight, then TV becoming common, then the landing on the Moon.

Neil Armstrong's Moonwalk
TV common in homes
World War II
World War I
Wright brothers' flight at Kitty Hawk

You may not know the date of the Wright brothers' flight at Kitty Hawk beach, but you can use clues to figure out whether it happened before or after the first Moonwalk.

By gathering information and making comparisons, we can put all five events in order. Because we have written records of when each event happened, we can also put them in order by using numbers. The Wright brothers' flight occurred in 1903, World War I lasted from 1914 to 1918, and World War II lasted from 1939 to 1945. Television came into our homes in the 1950s, and Neil Armstrong walked on the Moon in 1969.

Neil Armstrong's Moonwalk	1969
TV common in homes	1950s
World War II	1939–1945
World War I	1914–1918
Wright brothers' flight at Kitty Hawk	1903

Written records exist for only a tiny fraction of Earth's history. Understanding the rest of the history requires gathering data, comparing information, and making reasonable inferences.

The Geologic Time Scale

Scientists do not measure Earth's history in familiar units like days, months, and years. When we talk about Earth's history, time is measured in millions and billions of years. The geologic time scale divides geologic time into manageable chunks.

To develop a geologic time scale, geologists in the nineteenth century started by looking at fossils in different rock formations. Certain kinds of fossil shells were always found in layers older (lower) than layers with plants. Dinosaur fossils show up only in layers above layers with plant fossils. The geologists could use fossils to develop a relative time scale for the history of Earth. They inferred that species with shells had been living on Earth long before plants appeared. And dinosaurs appeared only after plants.

Study the modern time scale presented here. **Eon** is the largest division of geologic time. Eons are divided into smaller categories.

Eras are broad spans based on the types of animals on Earth at that time.

- **Paleozoic** means "ancient animal life." Life in the Paleozoic era included trilobites, corals, brachiopods, early fish, and early amphibians.
- **Mesozoic** means "middle animal life." It is known mainly as the age of dinosaurs.
- **Cenozoic** means "recent animal life." Mammals dominate the Cenozoic era.

Take Note

Both the Paleozoic era and the Mesozoic era ended in great extinctions of many species of organisms. What could that mean for the end of the Cenozoic?

The Relative Geologic Time Scale

The oldest time interval is at the bottom, and the youngest is at the top.

EON	ERA	PERIOD	EPOCH
Phanerozoic	Cenozoic	Quaternary	Holocene Pleistocene
		Tertiary	Pliocene Miocene Oligocene Eocene Paleocene
	Mesozoic	Cretaceous	Late Early
		Jurassic	Late Middle Early
		Triassic	Late Early
	Paleozoic	Permian	Late Early
		Pennsylvanian	Late Middle Early
		Mississippian	Late Early
		Devonian	Late Middle Early
		Silurian	Late Middle Early
		Ordovician	Late Middle Early
		Cambrian	Late Middle Early
Precambrian			

Periods subdivide eras. Periods are based on significant rock formations. For example, the Cretaceous period is named for English chalk beds of this age. (In Latin, *creta* means chalk.)

Epochs are subdivisions of periods. The Tertiary and Quaternary periods are divided into epochs. English geologist Sir Charles Lyell (1797–1875) came up with these subdivisions after he studied marine sedimentary rocks in France and Italy.

Index Fossils

William Smith (1769–1839) was an English surveyor and engineer. As a crew dug a new canal, he studied rocks in the canal's path. The exposed rocks puzzled him. Several exposures of limestone did not appear to be the same. Also, some layers were not horizontal, but were tilted and bent. Smith tried to figure out a way to make a **correlation** between rock layers at different sites.

Smith collected fossils from the limestone layers. He noticed that each different limestone layer had its own particular kinds of fossils. He could determine whether two samples of limestone were from the same layer by studying the fossils he found in each.

Some fossils showed up in several layers, but Smith found certain fossils present in only one layer. These unique **index fossils** could be used to identify a layer. Furthermore, he inferred that all layers containing a particular index fossil must have formed at the same time.

Correlating Rock Outcrops

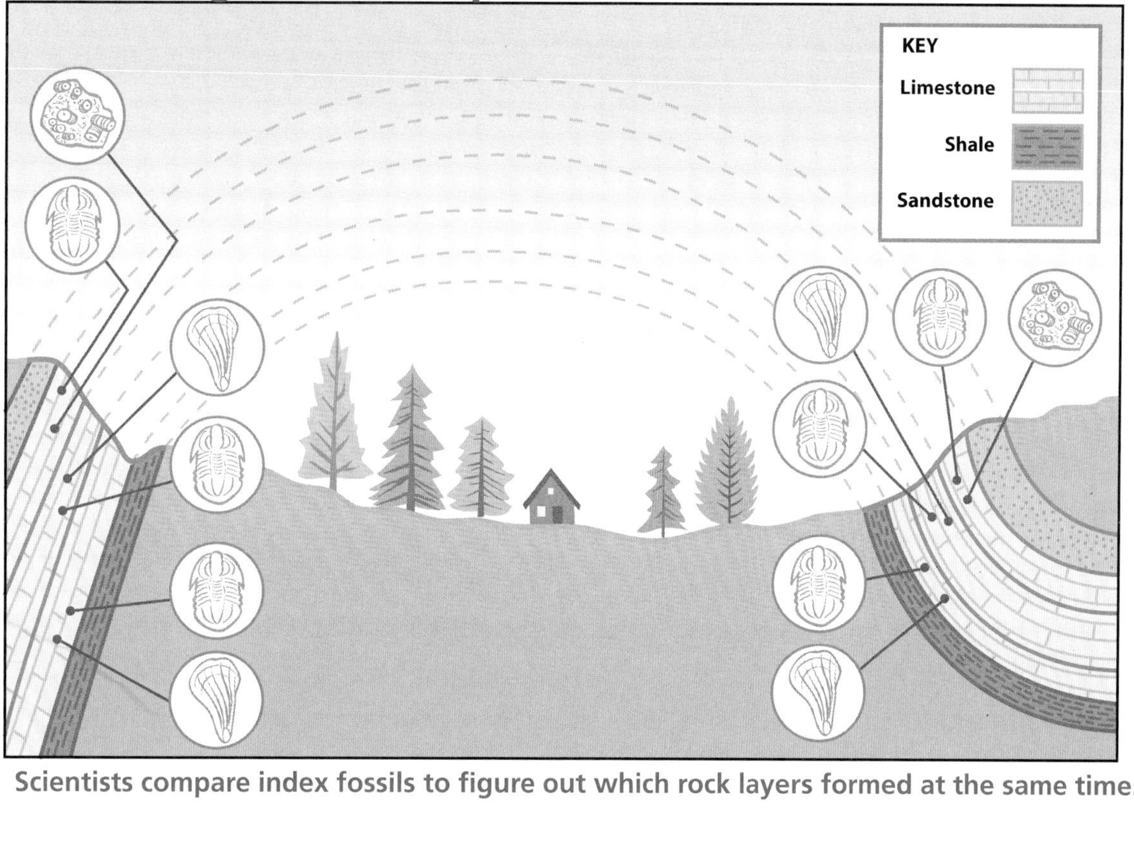

Scientists compare index fossils to figure out which rock layers formed at the same time.

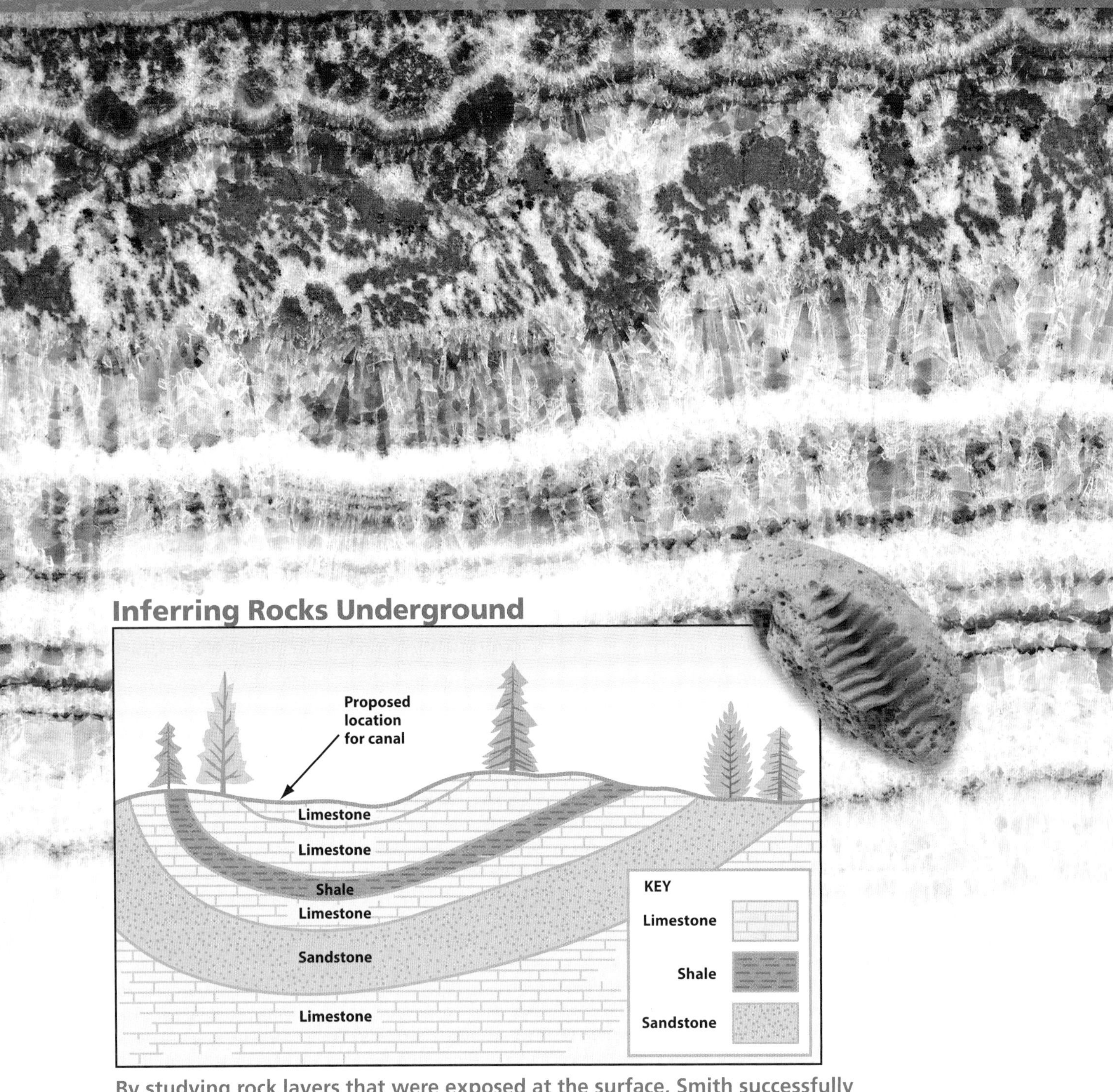

By studying rock layers that were exposed at the surface, Smith successfully predicted the order of rock layers underground.

Smith now had a way to determine the relative ages of the rock layers. He could also predict the order of rock layers below the surface. Smith drew a column of all the rocks below ground in an area. His drawing was accurate, even though there was no one site where all the rocks were exposed. These predictions were a great help in planning digging projects.

Smith's work established the disciplines of stratigraphy and the study of Earth's history. Geologists now use index fossils to relate rocks across great distances, even between continents.

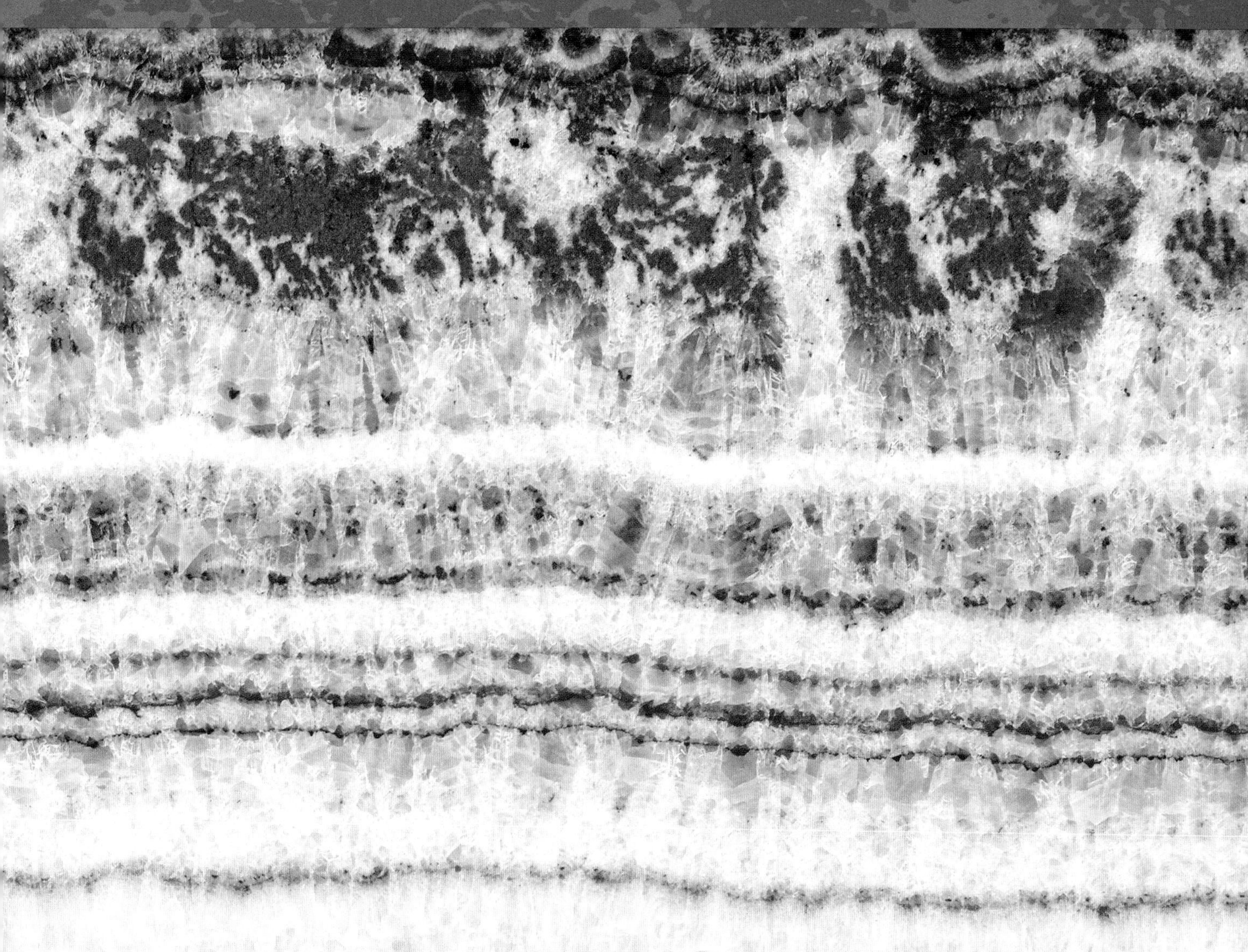

Stratigraphy is the study of rock layers. Stratigraphy exposes the earth history of a place going back many hundreds of millions of years.

Fossil Succession

If we begin at the present and look back into older and older rock layers, we soon reach a level with no human fossils. If we continue back through time, we reach levels with no fossils of flowering plants. Going even further back in time, we find rocks with no bird, no mammal, no reptile, no four-footed vertebrate, no insect, no land plant, no fish, and no shell fossils. We eventually get down to rock with no animal fossils at all, only bacteria. Finally, we reach a time when Earth rocks have no fossil evidence of life at all.

Three concepts are important in the study and use of fossils.

- Fossils represent the remains of once-living organisms.
- Most fossils are the remains of extinct organisms. They come from species that are no longer living on Earth.
- The kinds of fossils found in rocks of different ages are different because life on Earth has changed through time.

These concepts are summarized in the **law of fossil succession**.

- The kinds of animals and plants found as fossils change through time.
- When we find the same kinds of fossils in rocks at different locations, we know that the rocks are the same age.

This law is important to geologists who need to know the ages of the rocks they are studying. It is basically a time line of fossils.

New Forms of Life

How do scientists explain the changes in life-forms observed in the **fossil record**? Early explanations suggested natural disasters that destroyed life every so often. After each catastrophe, life began again.

In the mid-19th century, both Charles Darwin (1809–1882) and Alfred Russel Wallace (1823–1913) came up with another idea. They proposed that older forms of life produced new forms of life. Darwin's basic idea was that species evolve (change over time) or die out. Darwin's theory accounted for all the diversity of life, both living and fossil. This theory of evolution has been refined and modified as new information has been added. The new information supports Darwin's basic idea. Living things have changed through time, and older species are ancestors of newer ones.

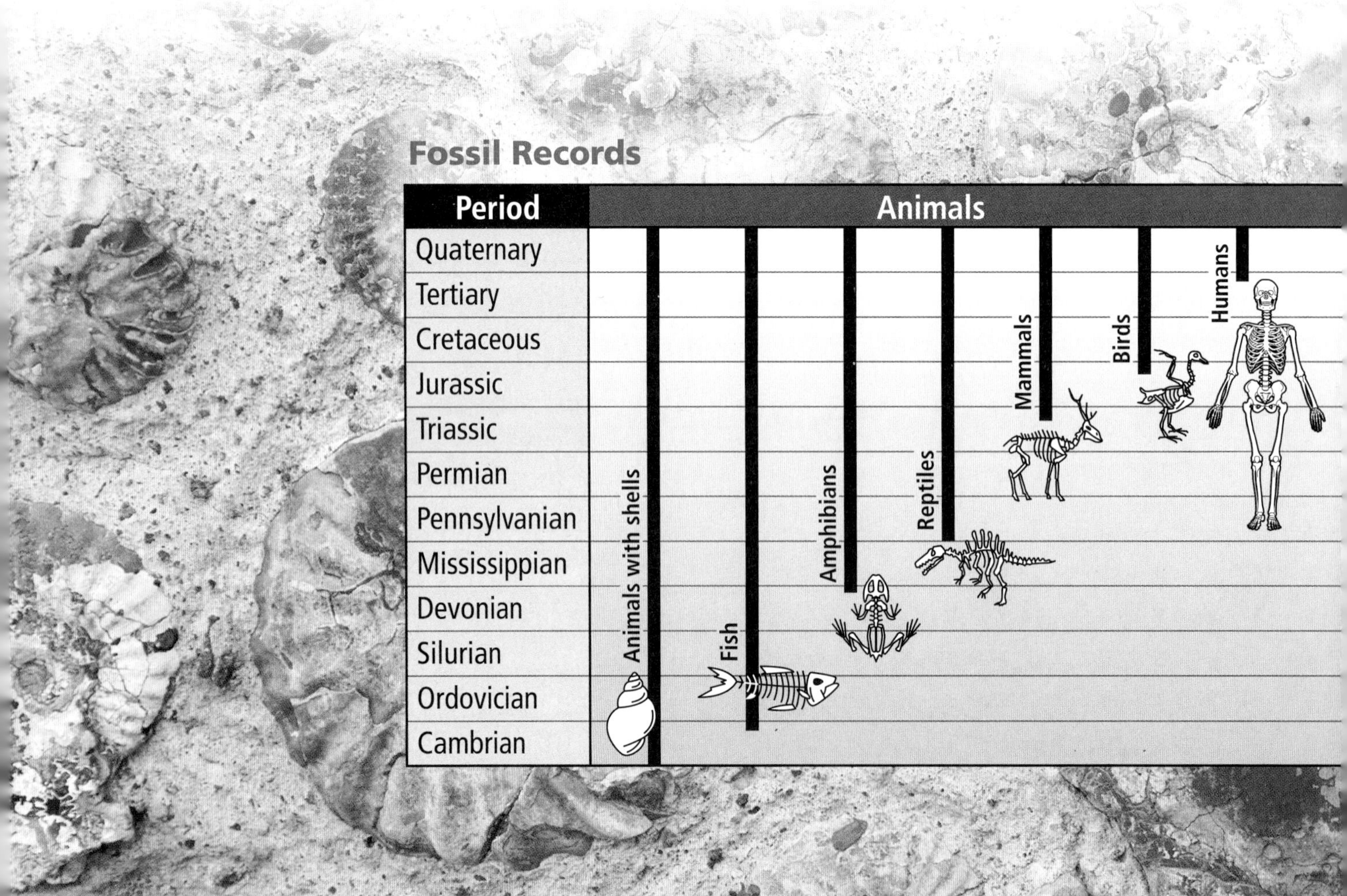

Fossils of a Known Age

Detailed study of many rocks from many places has produced a collection of index fossils. Index fossils allow geologists to "pinpoint" the age of a rock layer (within a few million years). An index fossil must meet certain criteria. The organism must

- have had a wide geographic range,
- have existed during a brief, well-known period of time,
- have been abundant, and
- be easy to identify.

One example of an index fossil is *Tetragraptus*. It was a small animal that floated in the early Ordovician sea that covered the Colorado Plateau. Because it floated, it had a wide distribution. It only lived during the Ordovician period. It is an important index fossil for Ordovician rocks. If someone handed you an unknown rock sample that contained the fossil *Tetragraptus*, you would know that the rock formed in the Ordovician period.

1 cm

Tetragraptus

Most fossils are found in sedimentary rock, often formed on sea floors. These ammonites were marine predators that became extinct along with the dinosaurs.

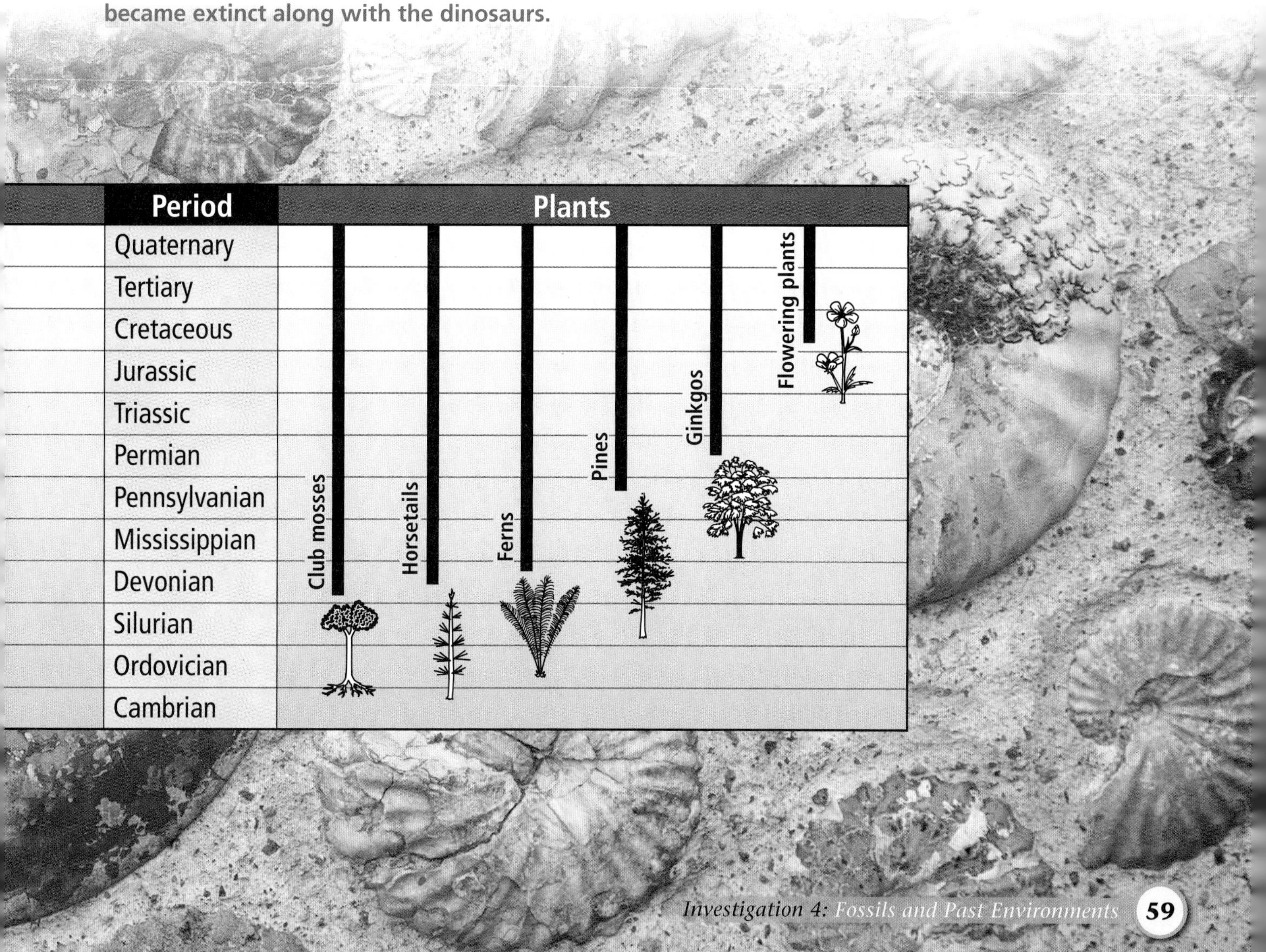

Period
Quaternary
Tertiary
Cretaceous
Jurassic
Triassic
Permian
Pennsylvanian
Mississippian
Devonian
Silurian
Ordovician
Cambrian

Plants
Club mosses
Horsetails
Ferns
Pines
Ginkgos
Flowering plants

The Absolute Time Scale

How did geologists get from a relative time scale to an absolute time scale? How did they determine these dates?

- Earth is about 4.6 billion years old.
- The oldest fossils are from rocks that are about 3.5 billion years old.
- The first abundant fossils with shells occur in rocks that are about 570 million years old.
- The last ice age ended 10,000 years ago.

Geologists and paleontologists in the 19th century had only crude ways of estimating how old Earth is. In 1896, they got a new tool for measuring Earth's age. Antoine-Henri Becquerel (1852–1908), a French physicist, found that the element uranium changed into lead through a process called natural radioactive decay. His discovery was the first step in finding a numeric age for rocks.

Not all elements undergo radioactive decay, only the ones that contain radioactive isotopes, such as uranium. Some isotopes decay rapidly, in seconds or days. Some can take millions of years. Only isotopes that take a long time to decay are useful to date rocks.

It takes 700 million years for half a uranium sample to turn into lead. This number is the isotope's half-life. By measuring the ratio of uranium to lead in a rock sample, geologists can calculate when the rock formed. At last, rocks could be dated absolutely!

Absolute Dating

Dating rocks is complicated and requires a well-equipped laboratory. Two commonly used methods are potassium-argon dating and carbon-14 dating.

Potassium is an element in many rock-forming minerals. Some of it decays into argon. Potassium-argon dating can be used on young rocks (a few thousand years old) and on rocks more than 2 billion years old.

At Dinosaur National Monument in Utah, a sloping rock formation contains fossils of hundreds of different organisms. Scientists can determine the age of sedimentary rocks by studying the fossils found in them.

Carbon-14 is better for dating a different group of objects. Carbon 14 constantly forms in Earth's upper atmosphere. It finds its way into all living things. After an organism dies, the carbon-14 in its tissues begins to decay. Carbon-14 is useful in dating only materials that were alive during the past 50,000 years.

Only **igneous rocks** can be dated absolutely. Geologists measure the potassium and argon in a rock's minerals and calculate when the rock became solid.

Dating Sedimentary Rocks

It is sometimes possible to accurately date rocks that are not igneous. The age of igneous rock that is next to sedimentary layers can be determined absolutely. This information can be combined with relative-age information to come up with an age for the sedimentary rocks.

Examine the diagram below. Which rocks formed first? Last?

Dating Rock Layers

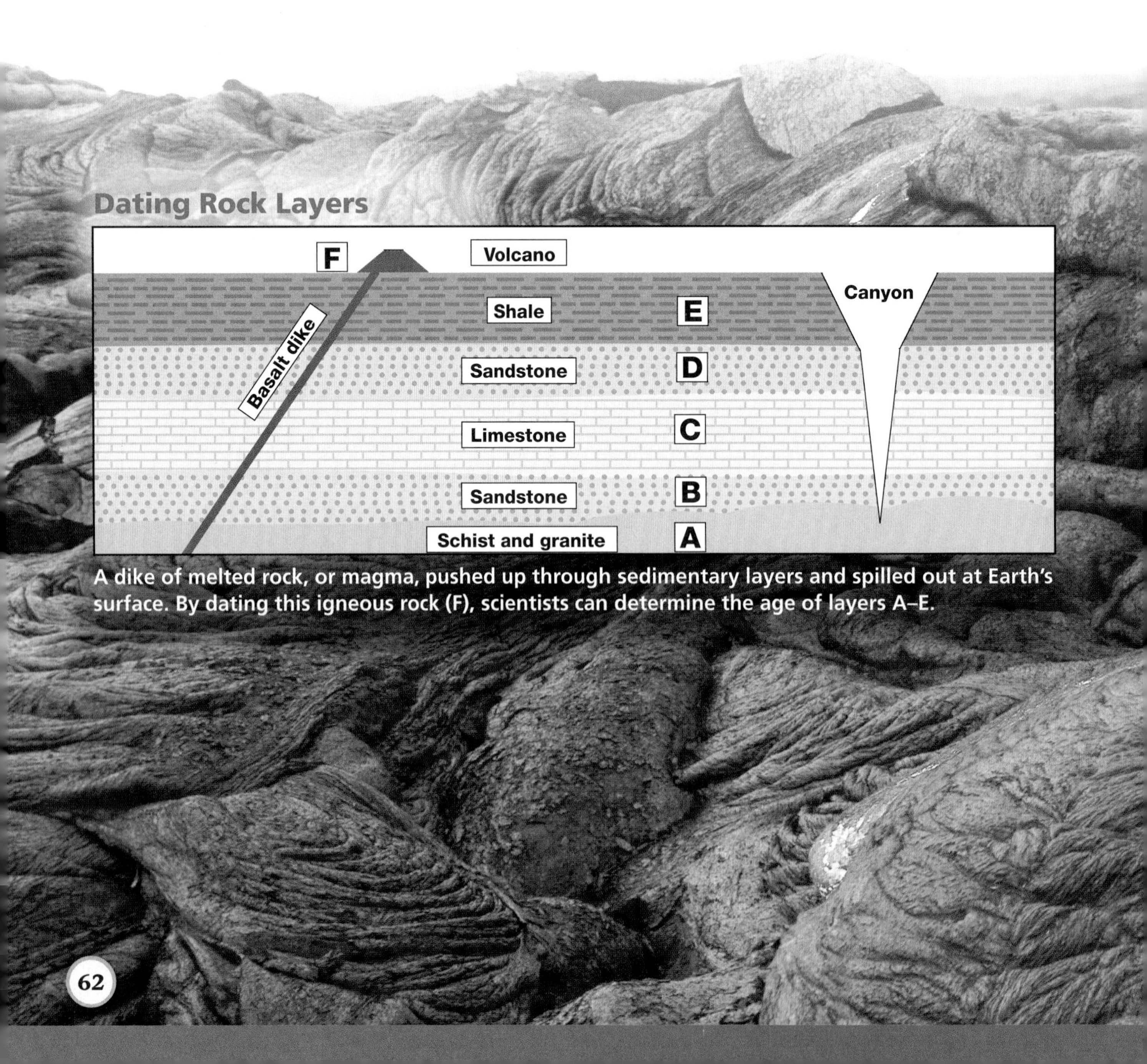

A dike of melted rock, or magma, pushed up through sedimentary layers and spilled out at Earth's surface. By dating this igneous rock (F), scientists can determine the age of layers A–E.

Geologists used potassium-argon dating on the basalt dike in the diagram. It is 23 million years old. This tells us that the sedimentary rocks, which had to be in place for the dike to form, are older than 23 million years.

By combining absolute age information with relative ages, geologists can figure out the dates of each new area of study on Earth.

This reading was adapted from Fossils, Rocks, and Time, US Department of the Interior/ US Geological Survey, USGPO Publication 1996-421-280.

Think Questions

1. **Explain several ways that fossils provide important data to geologists and stratigraphers.**
2. **Look at the chart showing when organisms appeared on Earth. What changes do you notice in the animals as you look from the most ancient times to the present?**
3. **Using the same chart, what changes do you notice in plants over time?**
4. **Give an explanation for the changes seen in organisms over time.**

The red hot rock is a lava flow that originated deep underground and has reached Earth's surface. As it cools, it will harden to form brand new basalt rock.

Floating on a Prehistoric Sea

For those who know how to read the language of geology, the rocks and landforms of the Grand Canyon contain the words that tell the story of its 1.7-billion-year history.

"One might imagine that [the Grand Canyon] was intended for the library of the gods; and so it was. The shelves are not for books, but form the stony leaves of one great book. He who would read the language of the universe may dig letters here and there, and with them spell words, and read, in a slow and imperfect way, but still so as to understand a little, the story of creation."

These words were written by John Wesley Powell in 1875. He was the first of many geologists to study the 1.7-billion-year history of the Grand Canyon.

Changes in Environment

The sediments that became the Grand Canyon's rocks were deposited by water and wind. We see these same processes of deposition today. Geologists use observations of current processes to interpret the evidence in Grand Canyon rocks. The rocks tell a vivid story of several seas, Sahara-like deserts, and swamps and tidal flats. These environments all existed at different times in this area.

Today, Indian Garden is a rare bit of greenery amid the stark Grand Canyon desert. It is hard to believe that this area was once the floor of a vast inland sea.

One sequence of rocks in the canyon tells the story of the advance and retreat of a great sea. The sequence includes the Tapeats Sandstone, the Bright Angel Shale, and the Muav Limestone. The sediments in these formations were deposited more than 500 million years ago (mya). They contain evidence of a very different environment than today's arid Colorado Plateau.

These three formations are at or beneath Indian Garden, an oasis on the Bright Angel Trail. Five hundred million years ago, there was no Grand Canyon. The Kaibab Limestone, the Coconino Sandstone, and the Redwall Limestone did not exist. A vast sea with crashing waves, tides, and coastal currents covered the area.

Sediment Sources

Imagine that you are floating in an inner tube on this prehistoric sea over the site of Indian Garden. Rivers erode the highlands and mountains to the east, carrying sediment westward. When the rivers enter the sea, their water slows down. Sand, the heaviest, most dense particles, settles near the shore. These sand deposits will become the Tapeats Sandstone. As you float on your tube, you can see the beach and enjoy the waves.

Sand Deposition

Running water erodes sediments. Sand is deposited at the beach and near shore.

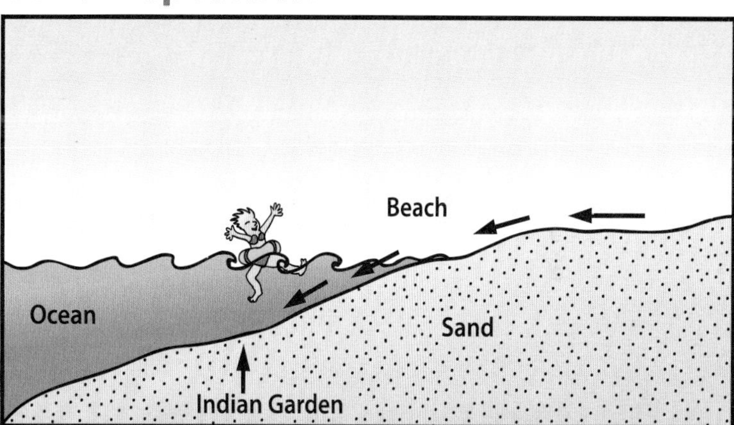

Clay and Silt Deposition

The ocean level rises. Lighter clay and silt particles are deposited farther from the beach, on top of the sand deposits.

Calcium Carbonate Deposition

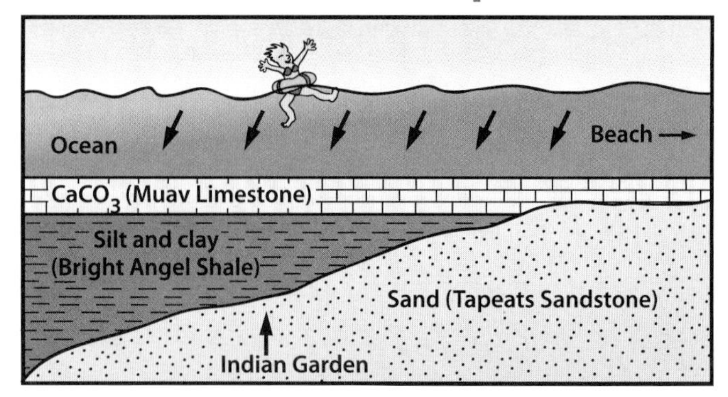

The ocean level continues to rise. The beach has moved so far "inland" that calcium carbonate ($CaCO^3$) is the main sediment deposited.

Millions of years pass. The sea is now much deeper under your inner tube, which is still positioned over Indian Garden. The beach moved farther away as the water rose. You are farther from shore, and the deeper water has less current. Small particles, silt and clay, settle on the seabed under you. A layer of silt and clay now covers the sand that was deposited millions of years ago when the sea was shallower. This silt and clay will later turn into the Bright Angel Shale.

The sea continues to rise. The shore is now far east of your location above Indian Garden. You can no longer see the beach. The water holds very little sand, silt, or clay. A layer of calcium carbonate begins to accumulate on the ocean bottom. It precipitates from the sea water, is released by organisms, or forms from the crushed remains of shells. The sea is full of life during this Cambrian period. Trilobites, mollusks, and brachiopods are plentiful. Their remains contribute to the calcium sediments that cover the older sediments, the future Bright Angel Shale and Tapeats Sandstone. These calcium-rich sediments will become the Muav Limestone.

More than 20,000 trilobite species are part of the fossil record. This type is found in both the Muav Limestone and the Bright Angel Shale.

Evidence of dinosaurs, lacking in the Grand Canyon itself, is found in nearby Tuba City, Arizona. Several footprints from at least four kinds of dinosaurs can be seen.

Missing Chapters

So there you float, meters away from the beach, still positioned above what will one day be Indian Garden. Don't worry, though. The water will recede, leaving the sedimentary layers high and dry.

More of the story appears in the rock layers of the Coconino and the Kaibab. But some chapters of the Colorado Plateau are missing in the Grand Canyon. No rocks in the Indian Garden area tell us about the time of the dinosaurs or the woolly mammoth. These creatures show up in the rocks of other areas. Rocks containing dinosaur tracks can be found near Tuba City, east of the Grand Canyon. But in the pages of the Tapeats, the Bright Angel, and the Muav rocks, we read the story of a time when Indian Garden was a sea inhabited by the interesting creatures that lived more than 500 mya.

Think Questions

1. **How does the sequence of rock layers help you tell the story of a place?**
2. **How do fossils help you tell the story of a place?**

Minerals, Crystals, and Rocks

What do all rocks have in common? Minerals. Minerals are the ingredients of all rocks. And minerals form crystals.

You already know some minerals, including calcite (calcium carbonate) and quartz (silicon dioxide). Calcite is the main mineral in limestone, and quartz is the main mineral in sandstone. Some rocks, like granite, are made of several minerals combined. Other rocks, like halite (rock salt) or calcite, are made of only one mineral.

Minerals are made of only one substance. Mineral properties include hardness, color, density, and cleavage. We use these properties to identify minerals.

Mineral Structure

Minerals are made of particles too small for us to see with our eyes alone, or even with a microscope. These particles are called **atoms**. When two or more atoms bond together, they form a molecule. For example, a quartz molecule is composed of two kinds of atoms, silicon and oxygen. Si is the symbol for silicon; O is the symbol for oxygen. The formula for a molecule of quartz is SiO_2.

Pure quartz, sometimes called rock crystal, forms as clear, colorless hexagonal structures. This specimen features iron pyrite (fool's gold) crystals on some of the tips.

A crystal is an orderly arrangement of atoms and molecules. Minerals become crystals when they change from a liquid to a solid. In the liquid state, either melted or dissolved, atoms and molecules can move around freely. When the solvent evaporates or the melted mineral cools, the mineral changes to a solid.

In a solid, atoms and molecules can no longer move past one another freely. But the positions of the atoms are not random. The atoms "click" into place, based on their properties. The structure formed by the atoms clicking into place is called a lattice. The lattice grows out in all directions, forming a crystal with a definite geometric shape.

Building a Lattice

You can explore lattices with a handful of pennies. Get at least 22 pennies. Put one penny in the middle of the table. Then place pennies around the edges of that penny, so that all the penny edges touch the pennies next to them. It should start to look something like the picture on this page. If you drew lines from the center of each penny to the ones next to it, you would see geometric shapes—triangles and hexagons.

Salt Crystal

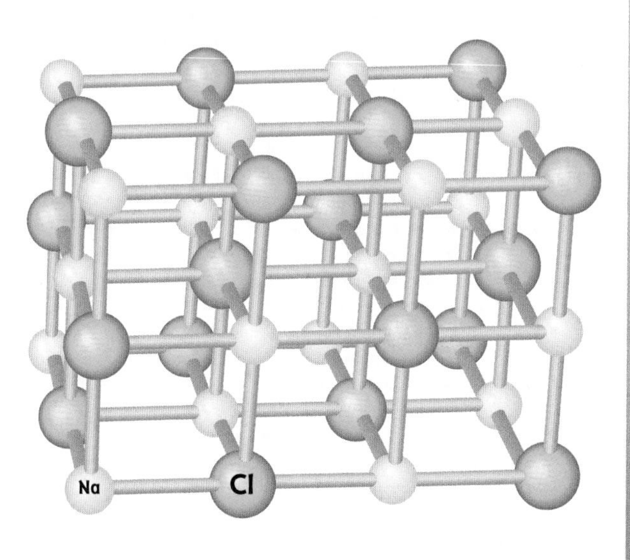

Table salt, or sodium chloride (chemical formula NaCl), is made of extended atomic structures that form crystals.

Model of a Crystal Lattice

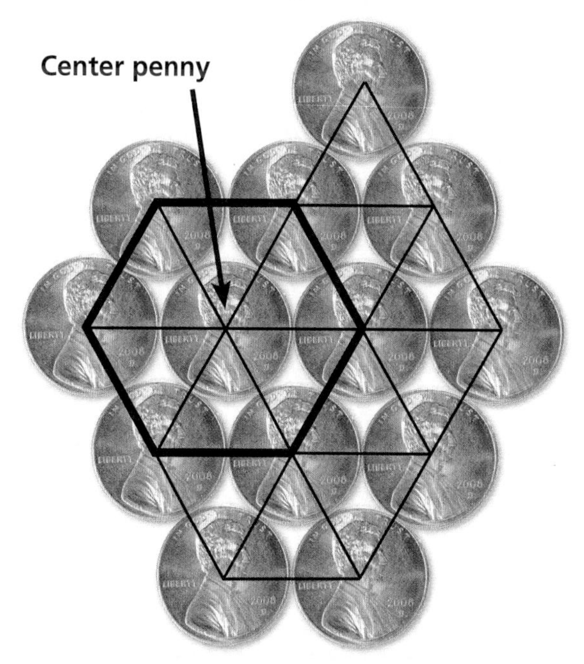

This arrangement of pennies models crystal formation as atoms form regular, repeating internal structures called lattices.

What happens as you continue to add pennies around the edge? You continue to build the same shapes. Your pennies model what happens when atoms and molecules form a crystal as molten rock cools.

Crystal Arrangements

Quartz crystals always grow with a hexagonal cross section. If the crystal in the photo on this page had enough time to grow in an ideal environment, it would look like this drawing of a six-sided crystal.

Scientists have identified six different crystal lattice systems. In these six systems, 32 shapes of crystals can grow.

Mathematicians have figured out that these are the only possible crystal arrangements. The shapes can be changed by temperature, pressure, and impurities in the crystal.

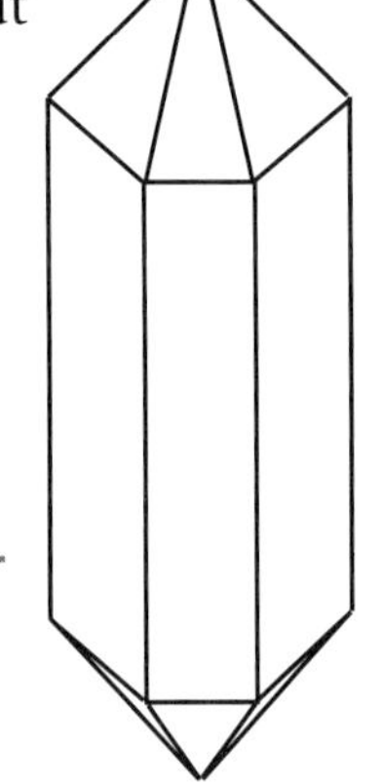

Amethyst and other quartzes form as six-sided crystals.

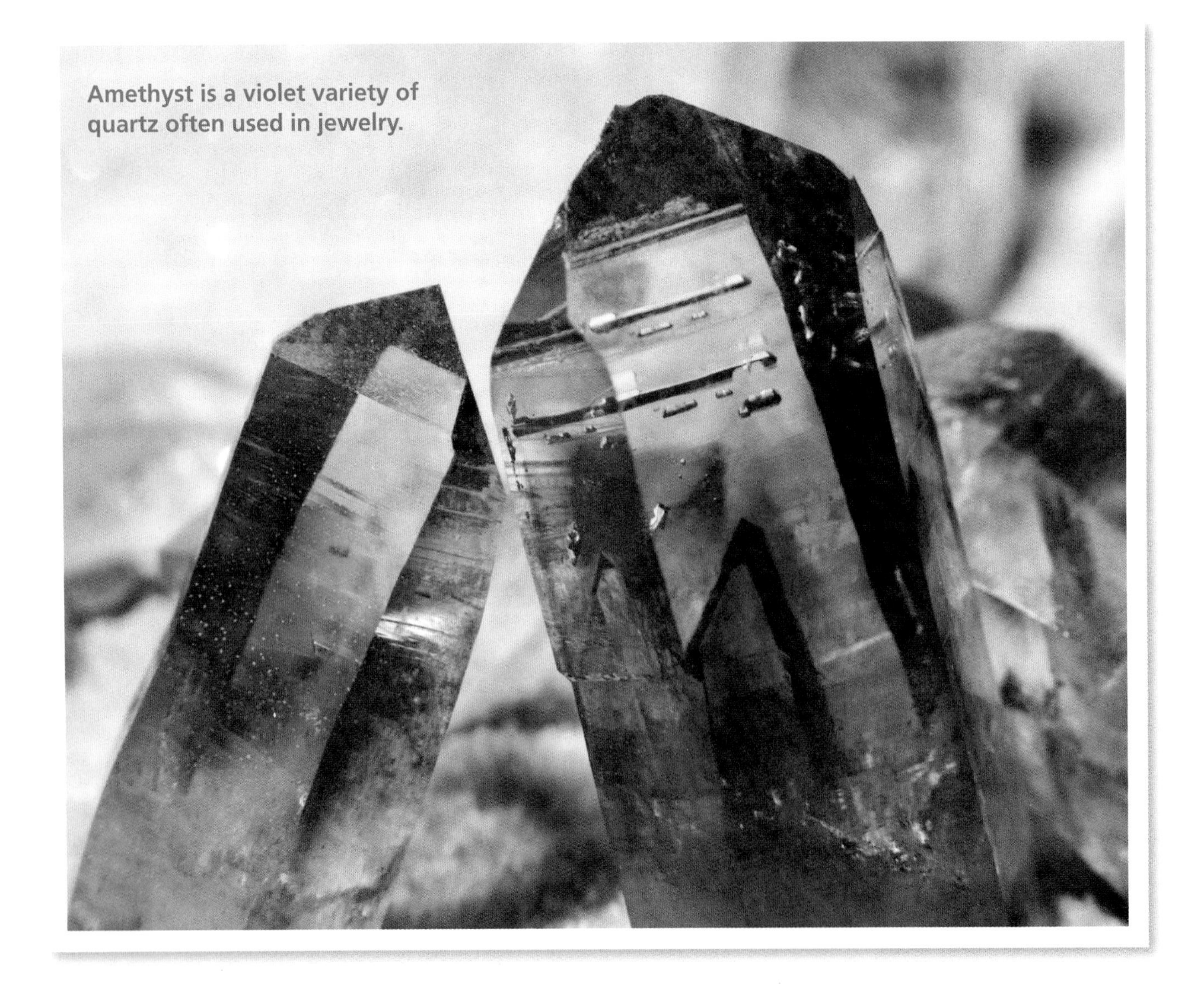

Amethyst is a violet variety of quartz often used in jewelry.

Kyanite is a mineral that forms "bladed" crystals, usually blue or green.

Conditions for Crystal Growth

It takes certain conditions for a perfect crystal to form. One condition is that the atoms and molecules must have complete freedom to move around when the material changes from liquid to solid. They can do this in molten rock (**magma**) or in a solution (like sea water).

Another condition that affects crystal formation is how fast the magma cools or the sea water evaporates. The slower the cooling or evaporation, the more time the molecules or atoms have to move into precise positions in the crystal lattice.

What happens when a solution of salt and water dries up in an evaporation dish? The atoms of sodium and chlorine that make up salt can move around freely in the water. As the water evaporates, it moves into the air as water vapor. When enough water evaporates, the sodium and chlorine can no longer stay in the solution. These atoms join to form a solid crystal of salt (sodium chloride, NaCl). The longer it takes for the water to evaporate, the more time the atoms have to get into position around the forming crystal. The crystal grows until all the water has evaporated.

If evaporation takes place rapidly, there is no time for atoms to move into the lattice of just one crystal, so crystallization may start in many locations at the same time. The result is a lot of small crystals instead of just one big crystal.

Mineral crystals form from magma or **lava** in a similar manner. As long as the magma is melted, the atoms or molecules are free to move around. Atoms or molecules of the same type are attracted to each other. If magma cools slowly in **intrusive** rock formation, the atoms have time to form large crystals as the igneous rock forms. But if lava reaches the surface and cools almost instantly in **extrusive** rock formation, there is no time for atoms or molecules to form large crystals. The crystals in fast-cooling lava may be microscopic.

What about Salol?

Salol is a synthetic material. Its chemical name is phenyl salicylate. Salol is used to manufacture plastics, suntan oils, waxes, and face creams.

Salol has properties that make it useful to us as we study rocks and minerals. It melts at a very low temperature and forms crystals when it cools.

Many factors come into play when salol crystallizes. It is hard to predict exactly how the crystals will form. It is fascinating to watch the crystals grow, especially if you observe them through a microscope. Natural minerals that originate as molten rock form in much the same way as the salol crystals.

Try This at Home!

You can grow your own crystals at home. Use the method described on the next page to grow crystals from a solution. Get adult help before starting this project.

Think Questions

1. **If you find an igneous rock that contains large crystals, what inferences can you make about how the rock formed?**
2. **Where would you expect to find crystals that formed from a solution?**

Rock candy is made of sugar crystals that grow together as a sugar solution slowly evaporates.

Crystals on a String

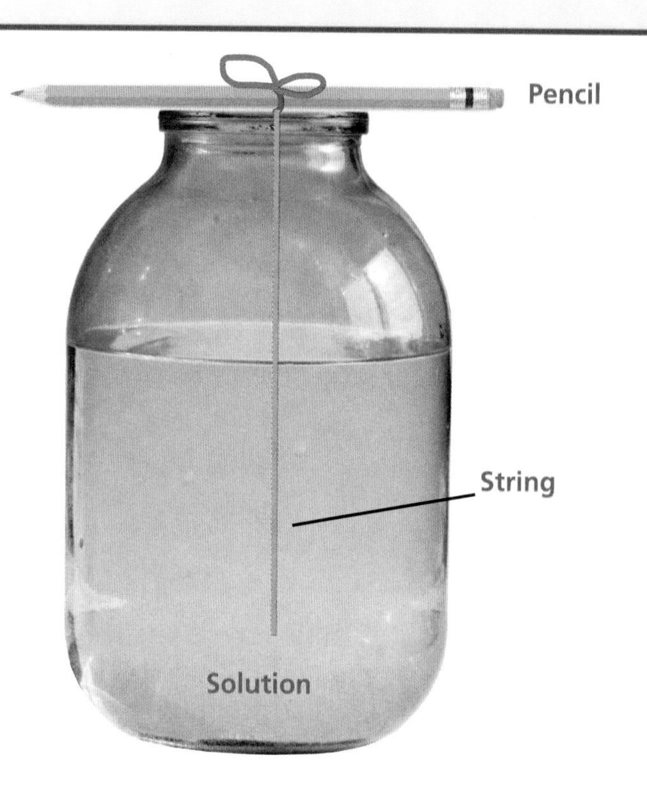

Materials

1 Glass jar, about 500 mL
• Hot tap water, 500 mL
1 Glass mixing container, 500 mL or larger
1 Measuring cup
1 Spoon or stirring stick
1 Piece of thin string, 15 cm long
1 Pencil
• One of these solids, 500 mL or more: kosher salt, Epsom salts, baking soda

Procedure

1. Measure 500 mL of very hot tap water into your mixing container.
2. Slowly add the solid material you have chosen to the hot water while stirring. Keep adding and stirring until the solution is saturated. When it is saturated, no more solid material will dissolve (disappear).
3. Pour the clear solution into the clean glass jar. You do not have to pour all the solution into the jar.
4. Tie the string to a pencil. Suspend the string in the center of the jar.
5. Let the solution rest for 15 minutes. Then swirl the jar a bit. Swirl it again 15 minutes later, then one more time 1 hour later.
6. Set the jar where it will not be disturbed.

Depending on the substance used, the crystals should begin to grow in an hour or so and continue to grow for several days.

NOTE: Old-fashioned rock-candy crystals can be grown using this same setup. Use sugar for the solid material, and grow the crystals on a wooden stick rather than on a string.

Once you have crystals growing, experiment with other substances. Investigate the different variables that affect crystal growth, such as temperature and rate of evaporation.

SAFETY NOTE: Use safe practices, such as wearing rubber gloves, when working with very hot tap water.

The History of the Theory of Plate Tectonics

Think about a time before seafloor maps and satellite photos of Earth, even before accurate global land maps.

The idea that Earth's outer layer is made of moving **plates** was not widely accepted until the 1960s. The theory of **plate tectonics** was probably not included in your grandparents' science textbooks. The theory is built on centuries of data and scientific development.

Geologic Puzzles

The origins of the theory go back to the first world maps in the late 1500s. These maps included most of Earth. After seeing the shapes of Africa and South America, some people wondered if the two continents were once connected. But how could that have happened? It took 300 years for scientists to come up with some ideas for how continents move.

As explorers traveled to the far reaches of Earth, they asked, How did fossils of sea creatures get on top of tall mountains? Is there a relationship between volcanoes and earthquakes? Were the continents once close together, making one big landform?

We now know that forces inside Earth are continually reshaping Earth's surface. As an oceanic volcano spews ash and cinders and sends rivers of lava to the ocean, it is creating new land.

Explaining the Puzzle

There were two main explanations for the mountaintop marine fossils. Some believed that global flooding raised the sea level above the highest peaks of the world. But otherwise, Earth's land had never changed. Others observed earthquakes and volcanic activity. They reasoned that processes inside Earth changed the surface, creating new hills and mountains.

James Hutton (1726–1797), a Scottish geologist, supported this second explanation. He observed that streams carried sediments away from his farm. So why had erosion not made the world into a perfectly round sphere? He decided that forces lifting sections of Earth's surface must balance out erosion. Hutton's theory required large amounts of heat energy from inside Earth and extremely long periods of time.

These were brilliant new ideas about Earth's history. But Hutton was a poor writer. Even the brightest scientific minds could not understand his written explanations.

After Hutton died, a close friend rewrote his book about geology. Eventually, scientists accepted Hutton's ideas of a changing planet. But understanding the evidence for plate tectonics was a long way off.

World maps in the 1550s began to give a more complete view of Earth. Some people began to wonder if South America and Africa had once been connected.

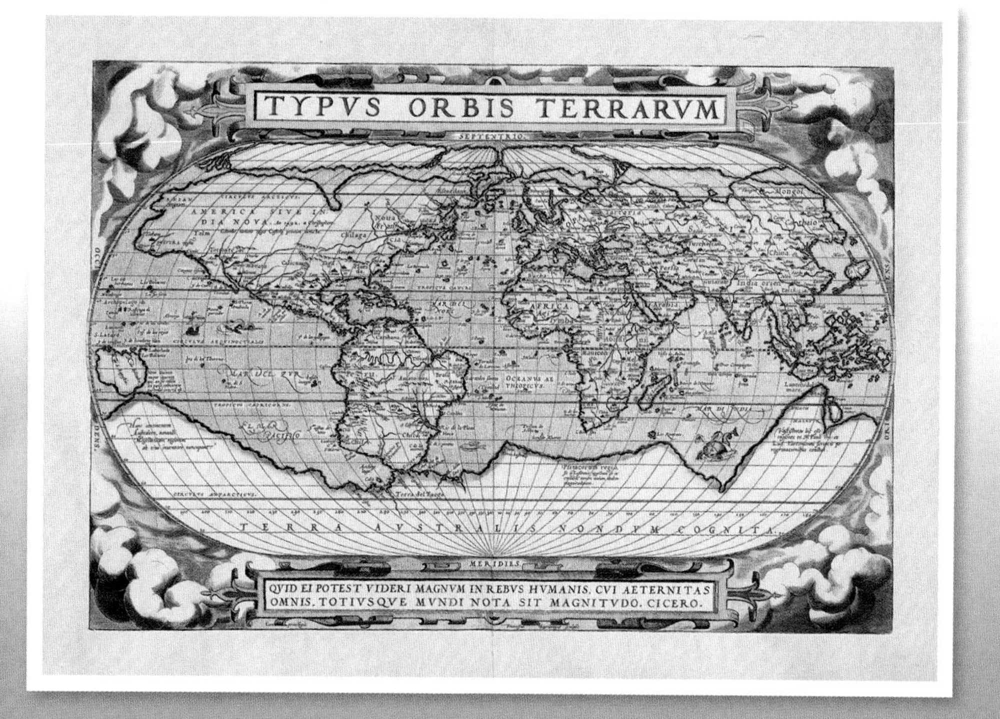

Putting the Pieces Together

The puzzle-like shapes of the continents intrigued Alfred Wegener (1880–1930), a German meteorologist. He was also interested in odd connections among fossils. For example, he found fossils of animals that once lived in tropical climates in areas that now have cold climates. He observed that the same plants and animals appeared as fossils in rocks of the same period on different sides of the ocean. The fossils included a freshwater reptile that was like a small crocodile found in Brazil and South Africa.

Can you imagine an entire community of reptiles traveling from Africa to South America? Neither could Wegener. Instead, he proposed a world where all the continents were connected as one huge continent. He called it Pangaea. He wrote of Pangaea as a land "where flora and fauna were able to mingle together before they were split apart." In the early 1900s, he published his idea of drifting continents, a new way of viewing Earth's history. But at the time, most scientists believed the continents were anchored in place. The continents might move up and down, but they certainly did not drift around the planet.

Earth's Landmasses in Ancient Position

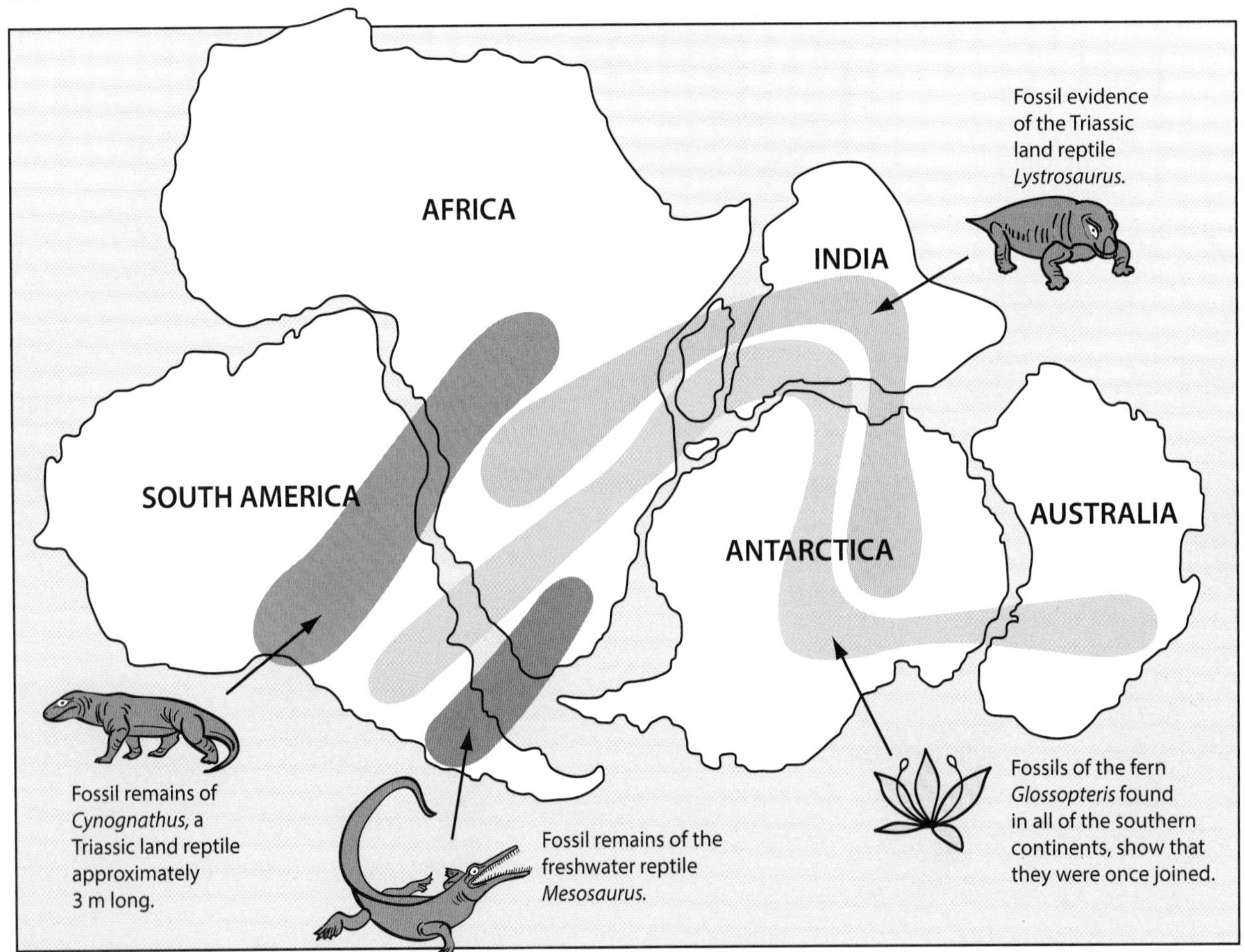

The color splashes show the possible patterns of fossil distribution before the continents split apart.

Earth Maps over Time

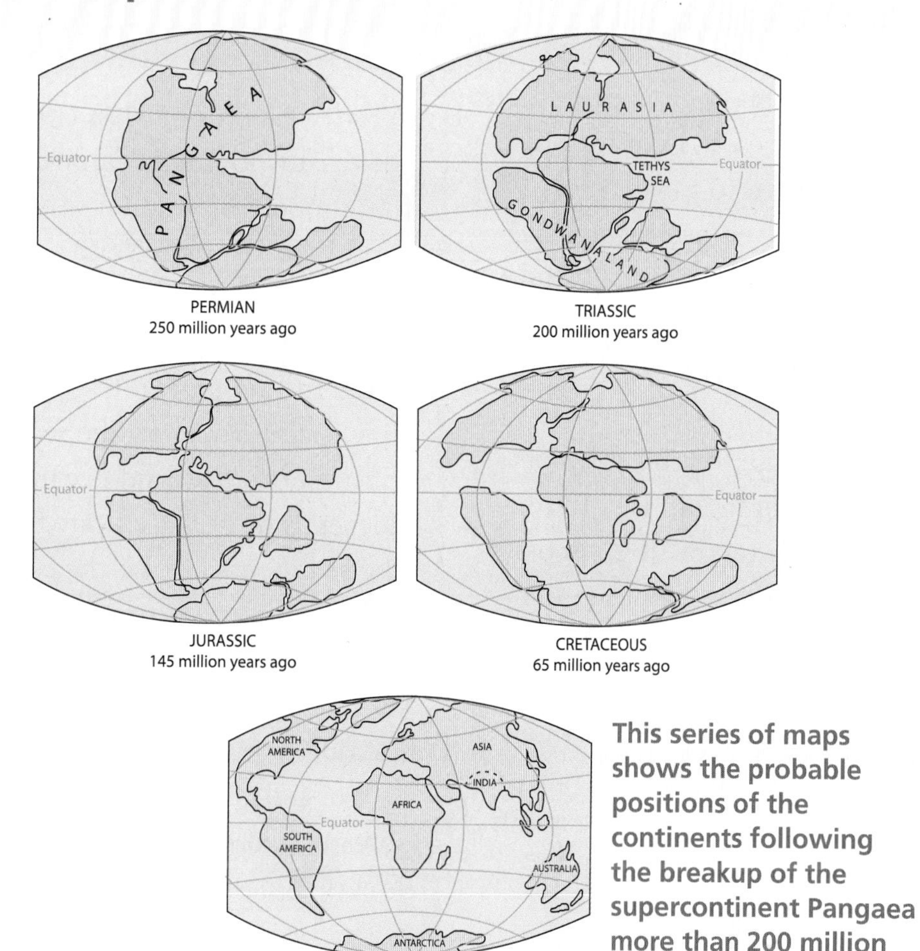

This series of maps shows the probable positions of the continents following the breakup of the supercontinent Pangaea more than 200 million years ago.

Resistance to Continental Drift

Scientists dismissed Wegener's ideas as physically impossible. In fact, Wegener could not explain what forces had moved large masses of solid rock over such great distances. How could it have happened? Wegener fought for his ideas of moving continents and a more dynamic planet until his death in 1930.

Scientists found it hard to ignore the fossil evidence for Wegener's theory. To get around this problem, geologists imagined land bridges crossing the ocean. Fossils of an ancient horse were found in France and Florida. So a land bridge was drawn across the Atlantic Ocean. A lost continent was an attempt to explain connections that spanned the Indian Ocean. These land bridges disappeared when technology allowed us to map the seafloor.

The Missing Pieces

In the 16th century, sailors used long ropes to measure the depth of the ocean. They found out that the seafloor is not as flat as people thought. Knowledge of the topography of the seafloor increased with the development of modern tools. In 1853, US Navy lieutenant Matthew Maury (1806–1873) published the first ocean-bottom chart. It showed evidence of underwater mountains in the central Atlantic Ocean. Survey ships laying the trans-Atlantic telegraph cable confirmed Maury's findings.

The first sonar systems were developed during World War I (1914–1918) to help locate German submarines. These systems recorded the time it took for sound to travel from the ship to the seafloor and back again. The ocean depth was calculated from these times. The data collected from this early sonar confirmed the existence of the Mid-Atlantic Ridge.

Sonar can also measure the thickness of sediments, an important tool in the study of the seafloor. In 1947, **seismologists** on the US research ship *Atlantis* measured the sediment layer in the Atlantic. Scientists had believed that the ocean was at least 4 billion years old. If it was that old, the sediment layer on the ocean bottom should have become very thick. But evidence from sonar readings showed that the layer was relatively thin. Why was that?

Seafloor Spreading

Harry Hammond Hess (1906–1969) was the geologist who finally came up with evidence that supported Wegener's theory of drifting continents. During World War II (1939–1945), Hess was captain of a transport ship equipped with the newest form of sonar. He decided to use the equipment to gather data all the time. He noticed that the sediment layer was thinnest near the mid-ocean ridge. It got thicker as he traveled away from the ridge. The sediments nearest the ridge had been deposited over less time than those farther from it. From this evidence, Hess and other scientists inferred that the crust nearer to the ridge was younger. This difference in age suggests that the Atlantic seafloor is spreading. It is pushing Africa and South America apart by about 5 centimeters (cm) a year.

Mid-Atlantic Ridge

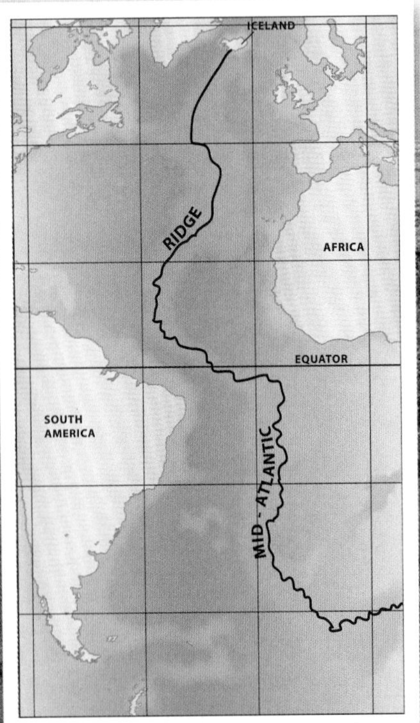

The Mid-Atlantic Ridge is part of the underwater mountain range that winds around the globe for about 70,000 km.

Today, we combine satellite images of underwater mountains with earthquake and volcano data to study the processes that shape Earth. These data are evidence that Earth's crust is composed of solid **tectonic plates**. They float on the fluid portion of the **mantle**. This evidence strongly supports the modern theory of moving crustal plates, called plate tectonics. It took courage and conviction for scientists like Wegener and Hess to defend and promote their unconventional ideas. Their new ideas changed the way everyone thinks about Earth.

Think Questions

1. What evidence caused Wegener to think the continents had been connected at one time?
2. Why did most geologists disagree with Wegener's ideas?
3. What are two pieces of evidence that scientists used to confirm the theory of plate tectonics?

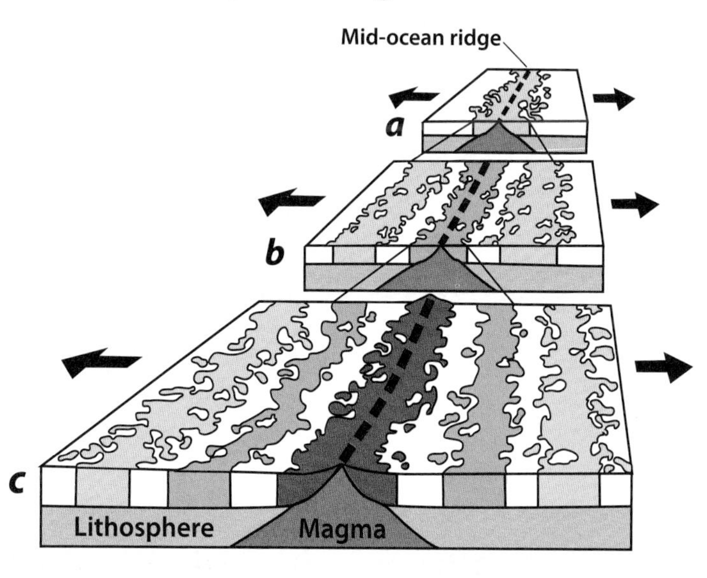

In seafloor spreading, new crust forms at the mid-ocean ridge and slowly moves sideways away from the ridge. Eventually this crust plunges back into the mantle at **trenches**.

The Mid-Atlantic Ridge—the boundary between the North American and Eurasian tectonic plates—slices through the center of Iceland. Rocky outcroppings mark where the plates are slowly pulling apart.

Historical Debates about a Dynamic Earth

Scientific theories are debated and challenged as new information emerges. Advances in technology help people collect new data. New evidence conflicts with some theories and strengthens others. Take a peek at the scientific theories about the story of Earth that people have used to answer two questions.

Why are ancient clamshells and other marine fossils often found on mountaintops?

Neptunists: Abraham Gottlob Werner (1750–1817) was a German geologist. He proposed that the land has not changed since it was created. Sea fossils on mountaintops are explained by the rise and fall of the ocean.

VS.

Plutonists: James Hutton (1726–1797) was a Scottish farmer, doctor, and naturalist. He proposed that land changed because of earthquakes and volcanoes. He also observed that hills and mountains could be eroded away. So he proposed that places must exist where hills and mountains are being created.

Are the forces that shape Earth sudden or gradual?

Catastrophism: Catastrophists believed that Earth was shaped by major events, such as floods and earthquakes, in a very short period of time.

VS.

Gradualism: The theory of gradualism, proposed by Hutton in 1795, suggests that Earth has been formed by very slow, gradual changes rather than abrupt catastrophic events.

Uniformitarianism: Sir Charles Lyell (1797–1875) took Hutton's ideas one step further. Lyell said that the processes that have shaped Earth are similar to Earth's processes today. The changes we see today are both sudden and gradual. Like Hutton, Lyell believed that gradual changes took place very slowly over long, long periods of time.

...ic Systems

...stantly changing. Weathering ...own mountains and deposits sedim...

7th

After billions of years, why isn't Earth's surface smooth? How do volcanoes and earthquakes change the surface? How do plate tectonics move continents?

Scientists describe Earth in terms of four interacting systems. The **geosphere** is the solid rocky surface and the interior of the planet. The **hydrosphere** is Earth's water, both in the ocean and on the land. The atmosphere is the air that surrounds Earth. The **biosphere** is all the living things on Earth. Let's take a closer look at the geosphere.

The names of Earth's four interconnected spheres—all shown here—are from the Greek words for Earth (*geo*), air (*atmo*), water (*hydro*), and life (*bio*).

The Geosphere

The geosphere has a thin, solid rock layer called the crust. Below that is a solid upper mantle and a massive fluid lower mantle. The inner and outer core are metallic. Geologists are most interested in the crust and the first 100 kilometers (km) or so of the solid mantle just under it. This region is called the **lithosphere**. Underneath the lithosphere is the **asthenosphere**, a fluid part of the mantle.

The lithosphere is the part that we stand on and that covers the seafloor. It seems that it should be one big, continuous covering on Earth, like the shell on an egg. But it is not. The lithosphere is broken into big slabs, like a hard-boiled egg with a broken shell. That's our picture of Earth today. This planet is molten rock covered with a bunch of solid plates of rock that fit together like puzzle pieces.

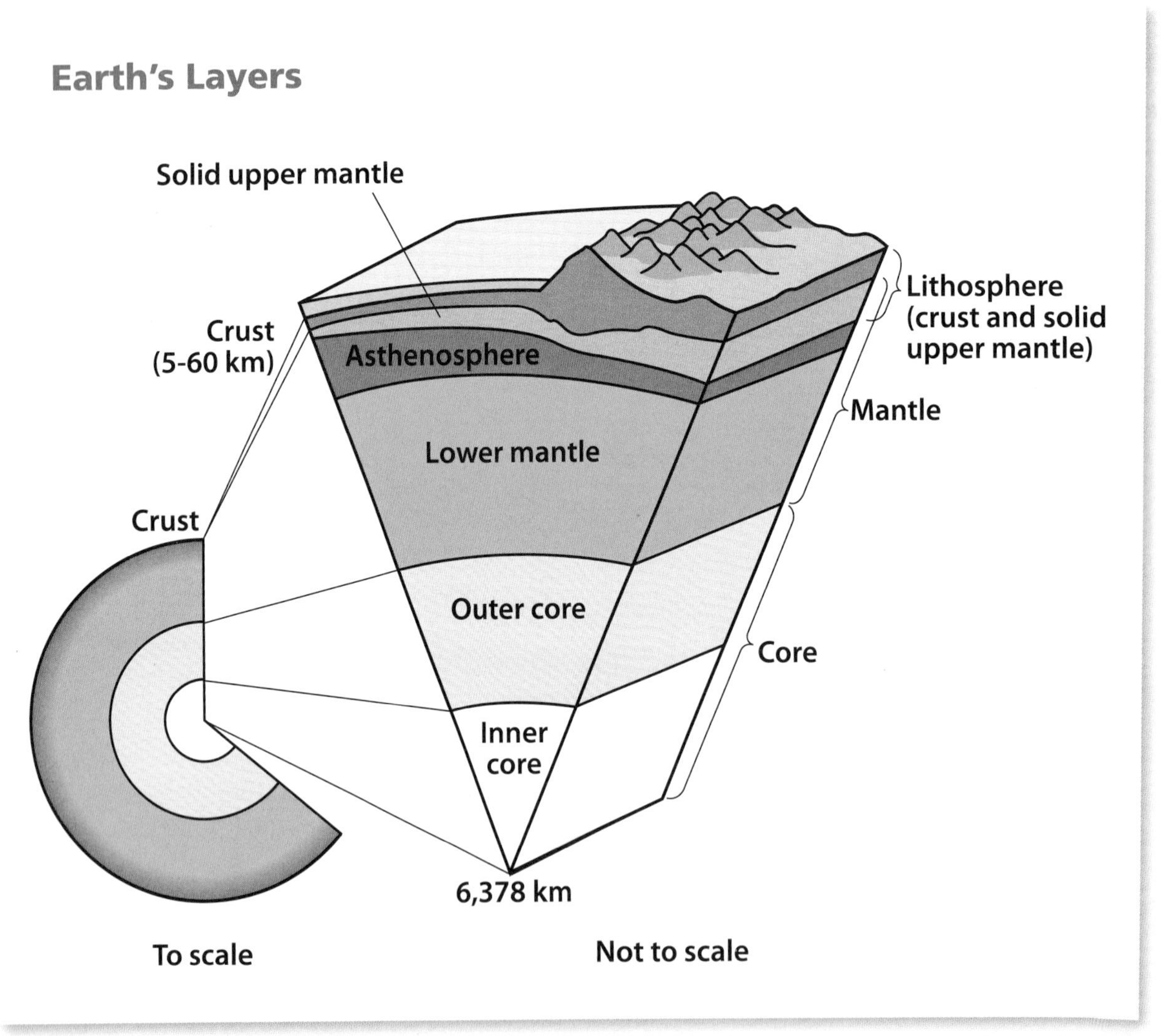

Geologists have learned that Earth is made up of three main layers: crust, mantle, and core. The layers vary by their composition and density.

Moving Plates

The lithospheric plates differ from the cracked shell in one important way. The lithospheric plates move around on Earth, while the pieces of an eggshell stay put. One of the larger plates is the North American Plate. All of Canada, most of the United States (except Hawaii, part of Alaska, and a slice of southern California), and most of Mexico wander across the surface of Earth together. Other large plates include the Pacific Plate, the African Plate, the Eurasian Plate, the Australian Plate, and the South American Plate.

So what makes the plates move around? Edges of solid, dense lithosphere sink into the softer, less dense asthenosphere. This **slab pull** causes plates to move apart. New rock forms between plates as magma rises from below. Circular movement in the mantle, called **convection**, occurs as the plates move around. These forces drive some plates away from each other, some plates toward each other, and some plates past each other. The San Andreas Fault is on the west coast of the United States. It marks where the North American Plate and the Pacific Plate are scraping past one another. The plates do not move fast, maybe 1–10 centimeters (cm) per year. But, as you know, geologists rarely think in units smaller than a million years. In a million years, a continent can move 10–100 km, and in 100 million years, 1,000–10,000 km. Now that's getting somewhere!

Plates

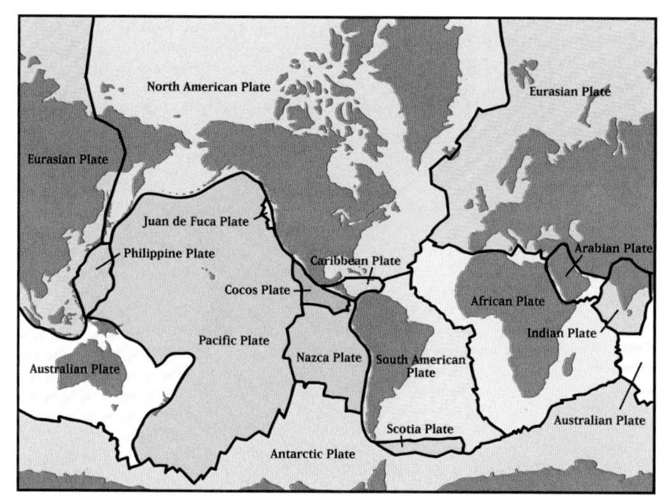

Earth's lithosphere is divided into a number of plates. Important geologic processes and events take place where they meet, at plate boundaries.

Slab Pull

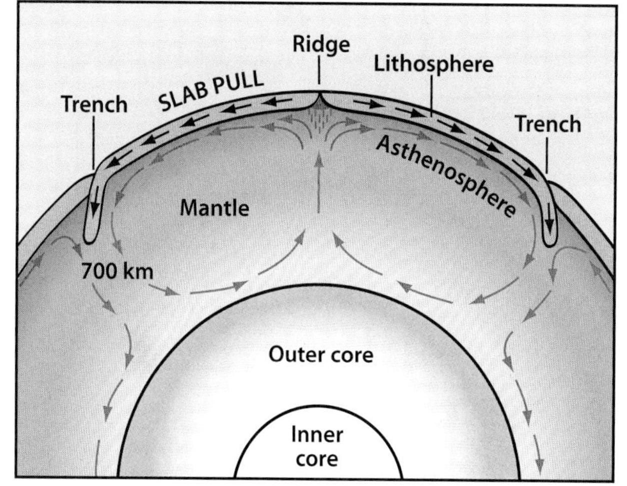

Slab pull occurs as lithospheric plates sink into the mantle, where the heated rock moves slowly in circular paths called convection currents.

Constructive and Destructive Processes

When two plates meet, or **converge**, one plate might slide under another. This is called **subduction**. When part of the sinking plate melts, the melted material can push up through the crust and onto the surface. The result is a volcano or a lava flow. Places with lots of volcanoes, like the southwest coast of Mexico and the west coast of South and North America, often indicate that an oceanic plate is sinking under a continental plate in a **subduction zone**. The Cascade Range from Canada to the middle of California has been created by volcanic activity.

Oceanic–Continental Convergence

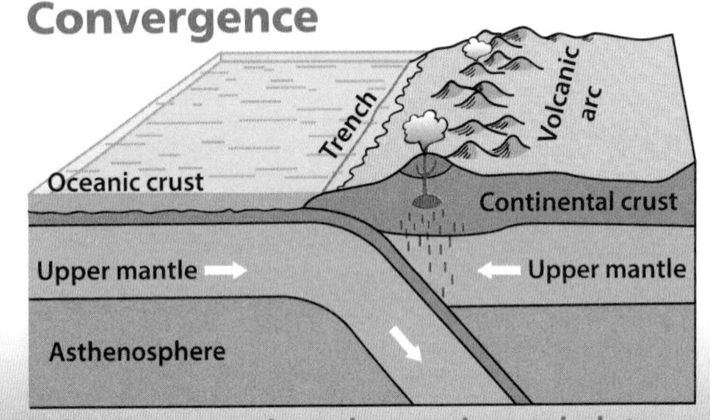

When an oceanic and a continental plate collide, the thinner oceanic crust slides under the thicker continental crust.

Sometimes when two continental plates converge, the continental crust gets folded and **uplifted**. The same thing happens when you push on one end of a small rug. The rug has to go someplace, so it forms hills and valleys. We see this happening in Asia today. India is converging with the Eurasian continent, rumpling up the landscape to create the Himalayas. These mountains get higher each year, about 6 cm per year.

Continental–Continental Convergence

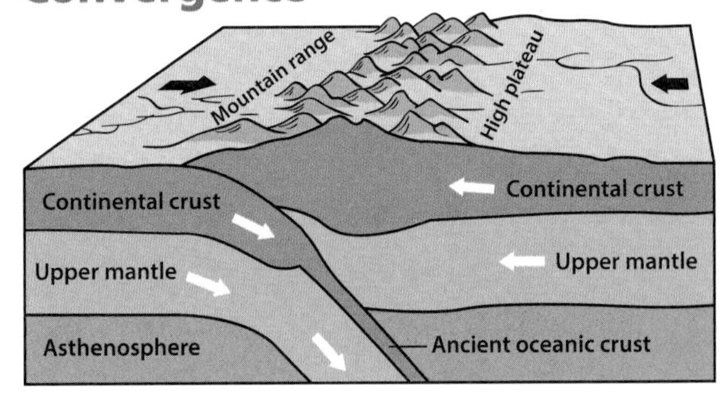

When two continental plates converge, the crust buckles and is pushed up into mountain ranges.

The Himalayas, already the world's highest mountains, are getting higher, rising about 6 cm a year as the Indian and Eurasian plates push against each other.

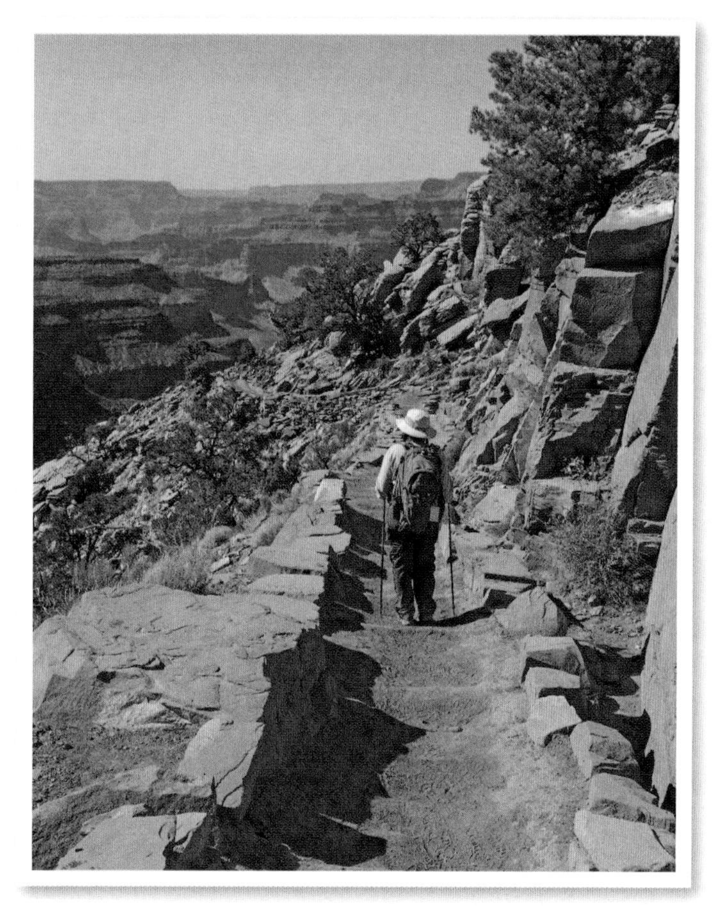

The South Kaibab Trail follows a ridge line into the Grand Canyon from the South Rim. It begins in the ledgy cliffs of Kaibab Limestone, one the highest, and youngest, rock formations in the area.

It is possible that something like this happened near what is now the Grand Canyon. Millions of years ago, a mountain range resulted when two plates converged. Then, the forces of wind and water broke the mountains down to dust. They swept the dust into the basin to form the sedimentary rocks of the Grand Canyon.

Earth is constantly creating surface and landforms as a result of several processes. Some processes are constructive. Mountain building results from plate convergence, with uplifting, **folding**, and volcanism. New crust forms where two plates are pulling apart. Sedimentation results from deposition. Destructive processes counterbalance the constructive processes. Gravity, wind, and water weather rocks. Erosion carries rocks away. Tectonic plates slide under other plates and melt.

Cracking the Kaibab Mystery

Perhaps we can use constructive and destructive Earth-shaping processes to answer questions about the Grand Canyon. Here we are, standing on the Kaibab Formation. Right under our feet are fossils of animals that once lived in a tropical sea. Yet the Kaibab Formation is more than 2,430 meters (m) above sea level.

Two possibilities spring to mind. Perhaps the sea used to be more than 2,430 m higher than it is today. The area where we are standing would have been underwater. Or perhaps the sediments we are standing on were 2,430 m lower, below sea level. Let's reason through these two possibilities.

No evidence on Earth suggests that there was ever an incredibly huge additional quantity of water covering almost all land. That eliminates the idea that the seas were 2,430 m higher a few hundred million years ago. What about the idea that the Kaibab Formation was deposited at or below sea level before being lifted up to its current height?

Index fossils and other evidence show that the Kaibab sediments were deposited near the end of the Paleozoic era, around 245 million years ago (mya). Other clues suggest that a major geologic event caused faulting, folding, and uplifting about 70 mya. What kind of event might have produced massive changes in the landforms? Maybe it was a convergence between plates. It could have been extreme magma activity under the North American Plate. The Rocky Mountains started rising at this time, too. The area that became the Colorado Plateau also started rising.

Faults and Folds

A **fault** is a place where Earth's crust is broken. The rocks on the two sides of the fault move past one another. Faults result when extreme force is applied to the crust to push it together or pull it apart. Rocks can break under the **strain** and slip past one another along a fault. The result is often an earthquake.

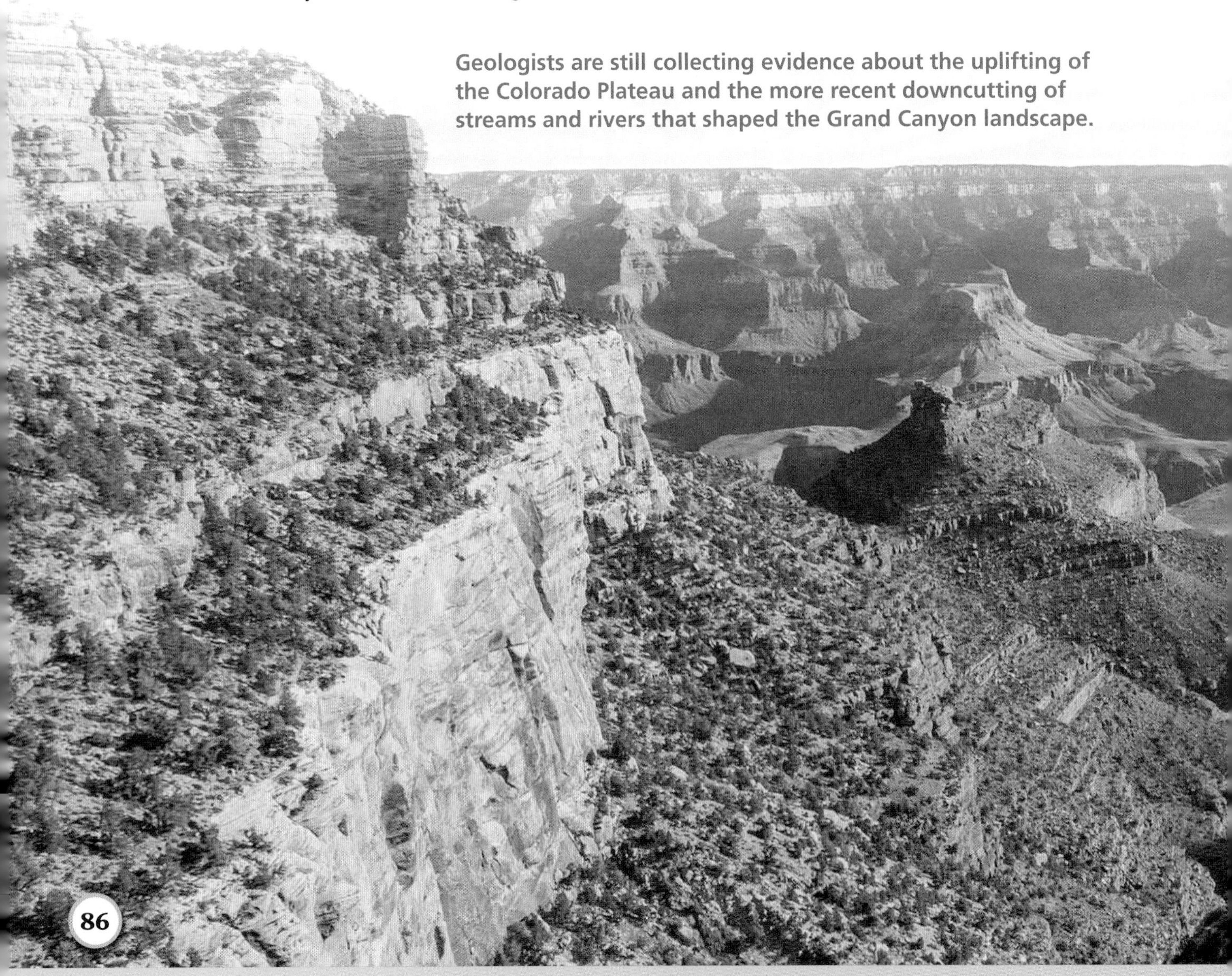

Geologists are still collecting evidence about the uplifting of the Colorado Plateau and the more recent downcutting of streams and rivers that shaped the Grand Canyon landscape.

Folds in the rocks indicate movement of Earth's crust. The crust has been compressed and lifted up. The existence of folds in the Colorado Plateau suggests that parts of the land were compressed during the elevation of the **plateau**.

Piecing together the history of the Colorado Plateau is a tough job. Part of the story is still a mystery. Geologists are sure the Kaibab Formation was deposited about 2,700 m lower in elevation than where it stands today. And the faulting and folding throughout the plateau suggest massive uplifting forces. But what events were the driving force to lift the Colorado Plateau? That is part of the fun of geology—there is always another mystery to solve.

Think Questions

1. **How can fossils found high in the mountains be explained?**
2. **If new crust is being formed at the mid-ocean ridge, why isn't the Earth getting larger?**
3. **What evidence indicates Earth is a dynamic (changing) system?**

Rock Transformations

How does one kind of rock change into another kind of rock? You observed what happened when you tried to crush some chocolate candy in a tube. It was tough work.

Miners seek out high quality marble. Slick gray and white swirls indicate that they've found the rock they are looking for. They will cut the rock into large pieces to construct a column, statue, or countertop. This beautiful rock took hundreds of millions of years to form. Let's explore that process.

What happened when you tried to crush some chocolate candy in a tube? It was tough work. Geologists think that something similar happens to rocks to change them. They infer that when rocks are put under pressure and warmed up, they change.

Geologists can put rocks under pressure in a laboratory and observe what happens.

But what they can do in the lab does not come close to what happens to a rock in the real world. It takes an *enormous* amount of heat and pressure to change a rock.

One thing to remember is that, if a rock gets hot enough, it melts. If it melts and cools, it becomes some type of igneous rock. To become a **metamorphic rock**, a rock gets hot, but not hot enough to melt.

Marble is a metamorphic rock that forms when limestone is subjected to the heat and pressure of metamorphism. This quarry is a source of marble for building material.

The Place for Change

Where is there just enough heat and pressure to make a metamorphic rock? That place is deep inside Earth's crust. Imagine you were a layer of sandstone and all your classmates were other layers of rock. If you were deposited first, and then all the rest of them were deposited on top of you, how would you feel? You might feel that you're under a lot of pressure, and a lot warmer, too.

Now think about a layer of rock that has been buried deep inside Earth's crust. It is under a lot of pressure because a lot of rock is piled on top of it. What about the heat? That is an interesting thing about pressure. Heat and pressure usually work together. As you go deeper into Earth, heat and pressure both rise. When you put pressure on something in a closed area, it warms up, too. If you could have measured the temperature of the candy in the tube, you would have found it got a bit higher. Not high enough to melt the candy, but warm enough to make it soft.

There is another source of heat. As the rock gets buried deeper and deeper, it gets nearer the hot mantle. Heat from the mantle can transfer to the buried rock, making it even warmer.

Yet another factor can increase the heat and pressure. Geologists call it strain. Remember how tectonic plates interact with each other? They can push together, pull apart, or slide past each other. Those movements can also put stress on rocks and change them.

Making a Metamorphic Rock

There is more than one recipe for making metamorphic rock, just as there is more than one way to make sedimentary and igneous rocks. Metamorphic rocks start out as some other type of rock, whether it is sedimentary, igneous, or another metamorphic rock. Normally, three things are required to change one type of rock into a metamorphic rock. These are lots of heat (but not enough to melt the rock), lots of pressure, and lots of time. If you had these three conditions and you could wait around for several million years, you might be able to watch a metamorphic rock forming.

Heat, pressure, and strain cause changes in rocks. Some or all of the atoms in the original rock are rearranged to form new minerals. The minerals in the metamorphic rock might be totally different from the minerals in the **source rock**, the rock you started with. New minerals, like garnets, might appear in the metamorphic rock.

Take Note

Review your metamorphic rock observations. How can you determine a metamorphic rock's source rock?

Heat and pressure can break chemical bonds and cause atoms to reorganize into new crystalline structures. Most garnets form as new minerals this way when rock metamorphoses.

Gneiss (above) and schist (right) show different bands of foliation.

Hardening. Sometimes the minerals in the source rock just become harder. For example, particles of the mineral quartz, which is often part of sandstone, can become more strongly cemented together. When you look at a piece of the metamorphic rock called quartzite, you can see that the quartz grains look sort of melted together. You will not see a matrix holding the grains together.

Foliation. Metamorphic rocks may also become foliated. **Foliation** happens when minerals like mica and feldspar in granite start moving around and form wavy layers of similar minerals. What the final metamorphic rock becomes depends on how much pressure is applied and how long it takes. Gneiss has wide bands of foliation. Schist has a finer texture with narrower bands.

Contact. Sometimes metamorphic rocks form when the source rock comes in contact with hot magma or lava. These rocks are baked and changed by the heat. Remember, they are still metamorphic rocks as long as they have not been heated enough to melt. Even underground **coal** fires can bake the surrounding rocks.

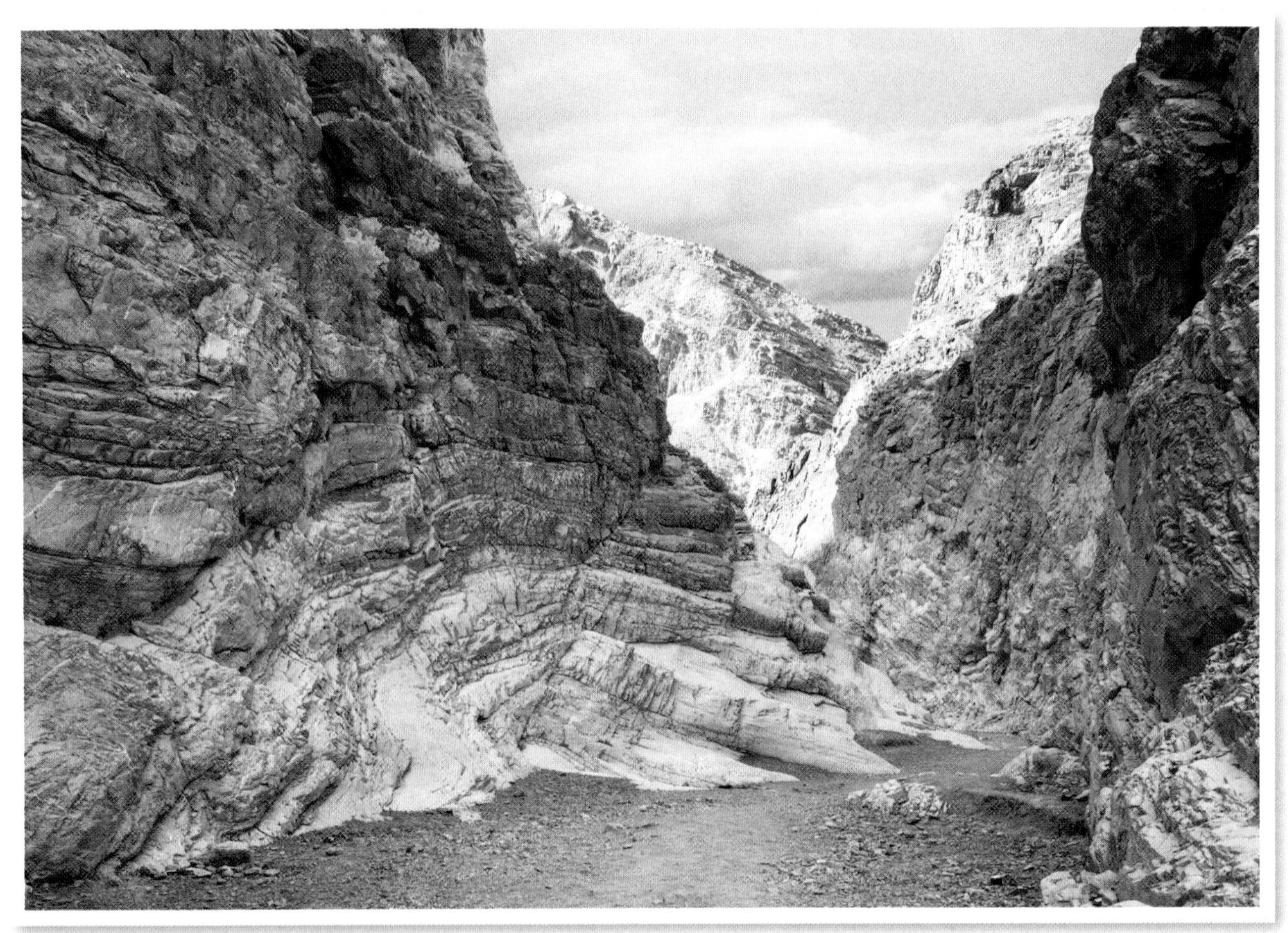

What evidence can you see that this rock is metamorphic?

Metamorphic Rock Types

Geologists have come up with a way to classify metamorphic rocks. You can see the relationship between rocks by looking at the metamorphic rocks chart in the Images and Data section of this book. Each row represents a different chemical composition. If you look across the row, all the rocks in that row started with similar chemical composition and are different only in how much they have changed. For each row, the source rock is on the left, the corresponding rock with low-grade metamorphism is in the middle, and the corresponding rock with high-grade metamorphism is on the right.

Think Questions

1. **What is *similar* about a source rock and the metamorphic rock that results from it?**
2. **What is *different* about a source rock and the metamorphic rock that results from it?**
3. **Where do new rocks come from?**

How One Rock Becomes Another Rock

It seems that rocks are constantly changing. Sometimes these changes happen rapidly, as in a volcanic eruption.

But most often the changes take thousands or millions of years. You would get pretty bored sitting and watching a rock, waiting for it to change.

As far as geologists know, every rock on Earth and in the crust has been something different in the past.

Let's review the processes that can change rocks. There are really only three that you need to keep in mind.

Melting and cooling. If any rock gets pushed deeply enough into the crust, extreme heat can melt the rock. It will become magma. The magma can cool in the crust or erupt onto the surface as lava. When magma or lava cool, they become igneous rock.

The Alps, formed by colliding tectonic plates, make a jagged profile across Europe. Scientists learn about mountain building by studying rocks and the forces that change them.

Weathering, erosion, and deposition. Physical and chemical weathering break down rocks. The sediments and chemicals from these rocks can be eroded and deposited in a basin. Over time they can become sedimentary rock.

Heat and pressure. Any rock that undergoes enough heat and pressure can be **transformed** into a metamorphic rock. It could happen when the rock is buried under kilometers of other rock. It might be caused by the incredible force of two continental plates converging. It could happen where magma rises to the surface. The magma's heat can change the surrounding rocks into metamorphic rocks.

Sedimentary rocks can become igneous rocks, or metamorphic rocks, or new sedimentary rocks. Igneous rocks can change into sedimentary rocks or metamorphic rocks, or. . . . You get the picture. These processes are the **rock cycle**. The materials in the rocks are constantly being recycled and rearranged to form new rock.

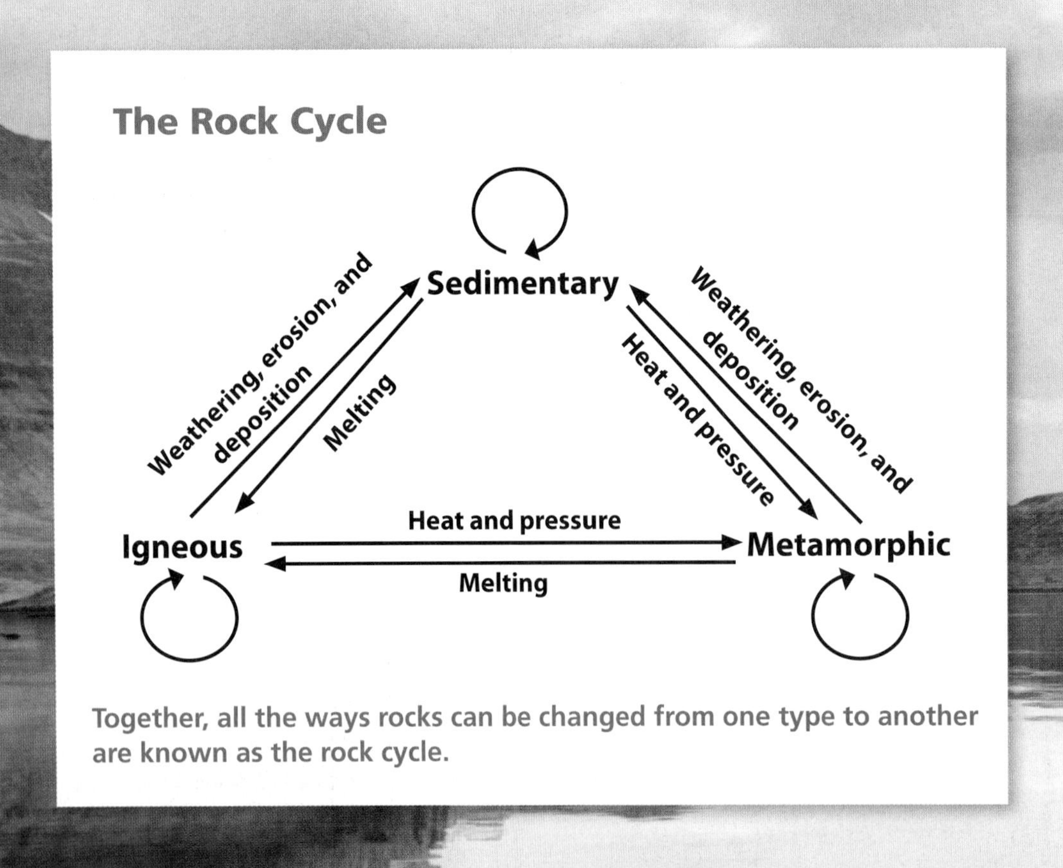

Together, all the ways rocks can be changed from one type to another are known as the rock cycle.

Take Note

Review your rock observations. What characteristics help determine the kind of rock (sedimentary, igneous, metamorphic)?

The Story of the Wrightwood Marble

Let's look at a rock that has taken several trips around the rock cycle. Wrightwood marble is a rock found in the San Andreas Fault Valley near Los Angeles, California. Its story started about 300 million years ago (mya).

Chapter 1: Melting and cooling. About 245 mya, a large pool of magma formed thousands of meters below Earth's surface. The magma contained calcium, magnesium, sodium, aluminum, iron, silicon, oxygen, and other elements.

Because magma is less dense than solid rock, the magma began rising toward the surface. The atoms in the magma began to organize into mineral crystals, like the crystals you saw in salol. After cooling for tens of thousands of years, the whole mass of magma crystallized into solid igneous rock. Let's follow some of the calcium in that igneous rock as we continue the story.

The weathering, erosion, and deposition that break down cliffs into pebbly beaches are part of the rock cycle.

Chapter 2: Weathering, erosion, and deposition. The movement of tectonic plates pushed up the igneous rock. After it reached the surface, the rock on top eroded away. Eventually all the igneous rock was exposed to sunlight and rain. For several million years, weathering broke the rock into sediments. The sedimentary particles containing the calcium ended up in rivers, on beaches, and in bays. Waves, currents, and tides eroded the small grains even more. Some of the calcium and other chemicals dissolved and flowed into the ocean. The calcium joined with oxygen and carbon to become calcium carbonate ($CaCO_3$). Clams and other sea creatures absorbed the calcium carbonate in their shells. When they died, their shells settled to the bottom. Some calcium carbonate settled to the bottom as ooze. The layer of calcium carbonate got thicker and thicker. The water was squeezed out, and over time a layer of limestone formed.

Chapter 3: Heat and pressure. Millions of years passed. Other rock layers were deposited on the limestone. The limestone got warmer and warmer as it was buried deeper. There was tremendous pressure from rocks above and from the Pacific Plate running into the North American Plate. This heat and pressure caused the calcium carbonate in the limestone to change. The rock became marble.

The Wrightwood marble cliff in southern California looks like this today.

If you want to make your own marble, start out with a piece of limestone and apply enough heat, pressure, and time. Put the same pressure on the rock as there would be under 25 kilometers (km) of rock or so. You would need to pile 200–240 cars on it to get enough pressure! You would need to heat it up to a few hundred degrees Celsius. Then wait for several million years while the atoms rearrange themselves. And *that* is how you make marble.

Chapter 4: Uplift. The calcium in the Wrightwood marble today was once part of an igneous rock on top of a mountain, and then part of a limestone buried deep below the ocean floor. When the limestone metamorphosed, this same calcium became part of the marble. When the marble was uplifted, it became part of a mountain again.

This open pit of marble was once an underground bed of limestone. Tremendous heat and pressure gradually changed it to the metamorphic rock so prized for building and sculpture.

What part of the rock cycle is represented by these lava flows? What type of rock is being formed?

The future. As millions of years go by, the marble will weather and erode. Water and wind will transport the sediments and minerals in the marble to a different place. These sediments and minerals will form different rock, and the whole process will continue.

The formation of every rock on Earth involves these or similar processes. And every rock is currently going through these processes and will become some other type of rock in the future. This process will continue as long as Earth exists.

Think Questions

1. **Develop a story of a rock cycle that includes sandstone, basalt, quartzite, and granite. The rocks do not need to be in the above order. Describe each step of the story. Include the type of rock, the location of the rock, and the processes that affect it.**
2. **Choose four or five rocks and make up a rock-cycle story. Explain how the material in the rocks changes and becomes part of a new rock.**
3. **What is a common type of rock in your community? Start with igneous rock and develop a rock-cycle story that ends with your local rock. Describe what you think might be in store for that rock during the next few million years.**

Geoscenario Introduction: Glaciers

You might be surprised to learn that many of the landforms we see on Earth today were formed by glaciers more than 10,000 years ago.

When you look at a map of North America, you are immediately drawn to the Great Lakes. Larger than the Grand Canyon, what sort of event could have created them? What happened to all the sediments removed to form them?

Many landforms that we see on Earth today were formed by glaciers more than 10,000 years ago. In this geoscenario, you'll learn about the evidence that glaciers leave behind paths and how they forever change Earth's surface. You'll learn to spot glacial evidence. You might even find some in your own community.

Glaciers once covered almost one-third of Earth's total land area. Today, glaciers are found mostly in and around the polar regions.

The Chugach Mountains along Alaska's south coast are covered by a series of connected glaciers, including the Bering Glacier, the largest and longest in North America.

The Great Lakes Area

Imagine taking a tour of the Great Lakes 12,000 years ago. The climate is warming slightly. The glaciers are slowly melting and retreating to the north. To the south is a **terminal moraine**. There the front of the ice sheet has stalled for a few hundred years. During that time, the ice sheet was moving forward at about the same speed as the front was melting. The ice sheet continues to move **glacial till** to the front of the glacier, adding to the terminal moraine. The moraine is over 100 meters (m) high in places. It is several hundred kilometers long, and several kilometers across.

Now go north to the front of the retreating glacier. The glacier towers more than 1 kilometer (km) above you. It stretches in both directions as far as you can see. That is a *lot* of ice! Between the moraine and the glacier are huge basins. They were gouged out by the ice sheet as it moved across this area for several thousand years. The basins are filled with meltwater.

The moraine, the mountains, and the towering glacier prevent the meltwater from escaping this area. These barriers create glacial lakes several times bigger than the Great Lakes. The Great Lakes formed about 11,000 years ago. That was after the glaciers finally retreated to the north. So the Great Lakes are one of the youngest geologic features on Earth.

The Great Lakes were formed by the movement of glaciers during the last Ice Age and filled with fresh glacial meltwater.

How Do Glaciers Form and Move?

Earth's climate is constantly changing. The last ice age, which lasted for 2 million years, had about 18 to 20 glacial periods. During these colder times, more snow fell in the winter than melted during the summer. Each winter added another layer of snow. As the layers piled up, the buried snow became ice.

The place that received the most snow was the Hudson Bay area in Canada. Over thousands of years, the ice sheet in that area was probably 3 or 4 km thick! The weight of that much ice caused the ice sheet to slowly flow in all directions. Part of the ice sheet moved as far south as the midwestern and northeastern United States.

Glaciers are always moving forward. As long as more snow falls than will melt in the summer, glaciers advance. When the climate warms, or when glaciers reach warmer regions, glaciers begin to melt as fast as they advance.

Sometimes a glacier melts faster than it moves forward. The front of the glacier retreats. During warmer **interglacial** periods, more snow melts during the summer than falls during the winter, and the glacier retreats. Earth is currently in an interglacial period. The only glaciers left are in the polar regions and at high altitudes.

Glacier Formation and Movement

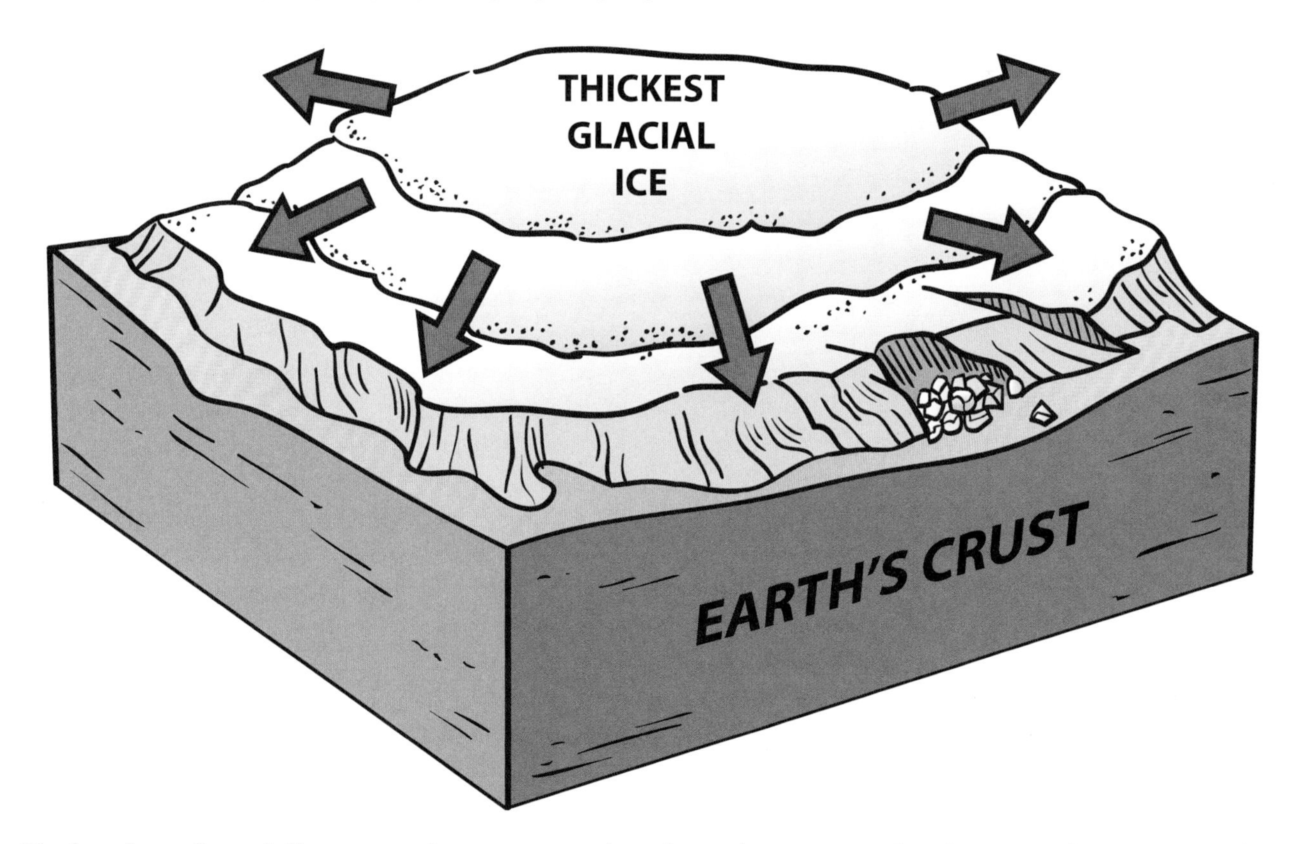

Glaciers form from fallen snow that compacts into huge ice masses. The downward pressure at the thickest regions causes a glacier to slowly move, or flow, downhill.

Have Glaciers Been Here?

When glaciers melt, they leave a lot of evidence behind. As glaciers move along, they scrape up and carry along a lot of earth materials, including large, **erratic** boulders. This poorly sorted material is called glacial till.

Melting glaciers, such as the one in the photo, act like a conveyor belt. They carry glacial till forward and deposit it where they melt. The deposit of glacial till at the front of a glacier is called a terminal moraine. If glaciers continue to melt at the same location for a few hundred years, they can build up a huge terminal moraine. One terminal moraine stretches along most of the New York–Pennsylvania border. Other terminal moraines are much smaller.

Glacial till also produces other formations. Sometimes a glacier deposits a long, narrow trail of till as it moves down the mountain and melts. Sometimes huge chunks of glacier, up to a few kilometers across, break off and are covered up by glacial till. When the ice melts, it leaves a large hole in the thick layer of glacial till. Water from the melted glacier fills the hole, producing a kettle lake.

As an ice sheet moves over an area, rocks embedded in the ice gouge out grooves and scratches in the bedrock. These scratches show where the glacier traveled and the direction of its movement.

Glacial till is sediment deposited by a retreating glacier. The mounds of till you can see at the glacier's edge may pile up to become landforms called moraines.

Geoscenario Specialists: Glaciers

This page describes some of the careers chosen by people who might study glaciers. You will choose one of these careers for your research in the class project.

A geologist interprets and explains information about rocks, rock formations, and landforms. Geologists can determine the date of geologic events by using absolute and relative dating.

This geologist looks for evidence of glacial till in a deposit of earth material.

A glaciologist studies the physical features of glaciers and how glaciers form and move. Glaciologists work with paleoclimatologists. They analyze ice cores and use glaciers to understand past climate profiles and make predictions about the future.

This glaciologist prepares an ice core to return to the lab.

A paleoclimatologist gathers data from seafloor and lakebed cores, glacial ice cores, coral samples, and other sources. Paleoclimatologists study past climates and make predictions about changes in our global climate.

This paleoclimatologist is collecting a seafloor core.

A climate policy adviser is responsible for understanding current scientific data about climate-change issues. Climate policy advisers recommend ways to minimize the negative effects of climate change.

This climate policy adviser shares data in a policy meeting.

Geoscenario Introduction: Coal

Whether you know it or not, you've used coal in your lifetime. Maybe it was burned to generate electricity for your home or your classroom; perhaps it fueled a factory that made your pants or calculator.

People have relied on coal as fuel for centuries. We now know that using coal can cause health and environmental problems.

How will coal use change as Earth's population (and demand for energy) grows? In this geoscenario, you'll learn about how coal formed, how people mine coal, and some of the issues for coal use in the future.

Coal Country, West Virginia

Coal is mined in many places in the United States. One of the oldest and largest coal mining areas is West Virginia. This entire state is in the Appalachian Mountains. The terrain is famous for its rugged mountains and hills. The forests of West Virginia are home to many animals, such as black bears, snakes, deer, moose, and mice.

Coal is a sedimentary rock that forms very slowly from the remains of plants that lived hundreds of millions of years ago, usually in a swamp environment.

During the Industrial Revolution, the demand for coal soared. It is an excellent fuel that was burned to power steamships and steam locomotives. Residents of West Virginia (Native Americans and settlers from Europe) had used coal from the mountains for fuel for a long time. Now they saw business opportunities. Coal mines quickly increased in size and number. The coal industry of West Virginia was born.

The Formation of Coal

Coal is an excellent fuel because it is energy dense. A piece of coal releases far more energy as it burns than the same mass of wood. The key to its stored energy lies in its formation.

Coal formation takes hundreds of millions of years. It starts with ancient plants. Plants convert energy from the Sun into molecules that are stored in the plants. Most coal deposits are formed from plants that lived and died around 300 million years ago, before the time of the dinosaurs.

At that time, dead plant material fell or washed into swamps. Over millions of years, sediments settled on the plant material. The plants disappeared under water and dirt. In these conditions, plants are slow to decompose. With enough time, heat, and pressure, partially decomposed plant material called **peat** can turn into coal. Coal is a rock that can be burned to release energy. Lignite and bituminous are sedimentary rocks, and anthracite is a metamorphic rock. Burning coal releases the energy stored by those plants millions of years ago.

Coal Formation

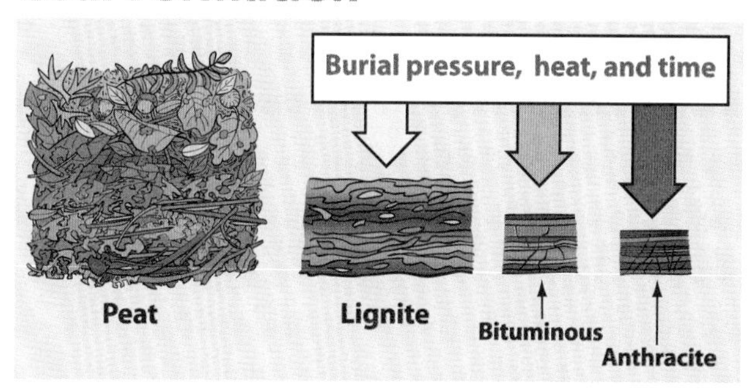

Coal starts developing from peat, and can form three different stages: lignite, bituminous, and anthracite. At each stage, the coal gets harder and contains more stored energy.

Moving Coal

Coal is found below ground. Sometimes it is just below the surface. It is often so deep, however, that people need a mine to get to it. Once the coal is dug up, railroads take it across country to customers.

The original Chesapeake and Ohio (C&O) Railroad line was built along the New River Gorge between 1869 and 1872. The railroad transformed an isolated land of small farms. The area became busy with logging and coal-mining towns. These towns supplied the natural resources that fueled the US industrial revolution.

The C&O Railroad was built mostly by thousands of Irish Catholic immigrants and freed African Americans. The work took 3 years of digging, grading the rail bed, blasting tunnels, building bridges, and laying tracks. The work was done with hand tools and explosives. Horses and mules helped with the heaviest loads.

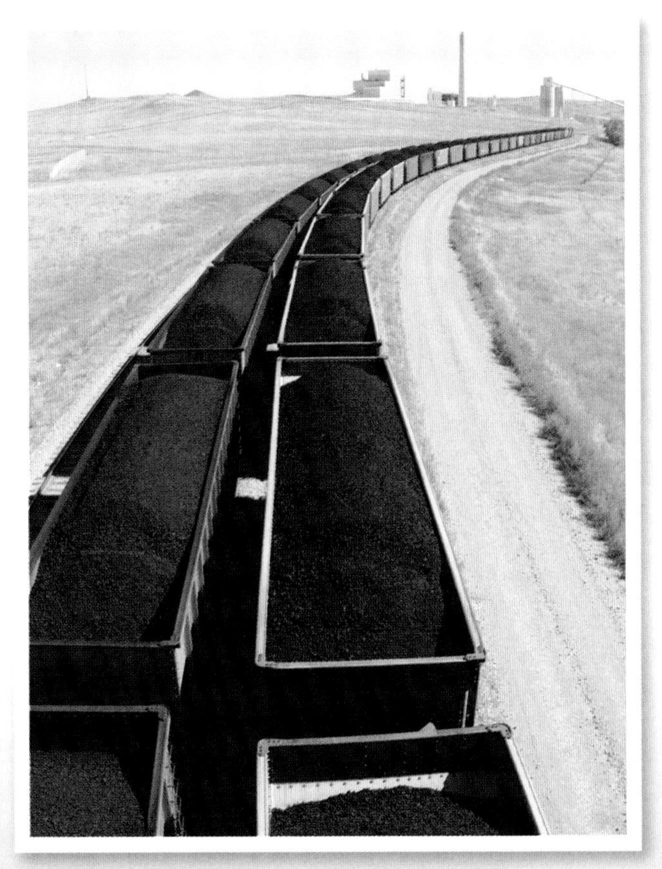

Coal travels from mines to power plants by train and truck. A typical coal train hauls fuel in 100 or more freight cars called hoppers.

Impact of Coal Mining on the Community

The coal mines are important to the economic well-being of West Virginians. However, coal mines come with costs, such as environmental and safety risks.

A mining accident that kills three or more miners is called a mine disaster. The New River Gorge was the site of three major mine disasters. However, most miners' deaths have occurred one at a time in roof falls and machinery accidents. Since 1883, 21,000 West Virginia coal miners have died.

Getting to coal deep underground requires heavy mining. One mining technique removes the entire mountaintop. This level of disturbance leads to problems with erosion and runoff. Water runoff from mining sites can be highly acidic. It can affect groundwater and drinking water.

Power plants burn coal to generate electricity for the community. Burning coal presents a number of problems. The process releases ash and carbon dioxide, among other substances, into the air. Ash pollutes the local area. Carbon dioxide moves into the atmosphere, where it is a **greenhouse gas**. These gases affect most of Earth, even far from the mines. There is strong scientific evidence that human-related carbon dioxide emissions cause climate change.

About one third of the electricity used in the United States is generated in coal-fired power plants.

Geoscenario Specialists: Coal

This page describes some of the careers chosen by people who might study coal. You will choose one of these careers for your research in the class project.

A geologist interprets and explains information about rocks, rock formations, and landforms. Geologists use absolute and relative dating to determine the date of geologic events.

This geologist examines coal samples in West Virginia.

A geographer studies the physical features of Earth and its atmosphere. Geographers relate these features to human activity. They find information about distribution of populations, resources, land use, and industries by interviewing people and accessing records.

This geographer uses computer modeling to track land use.

A coal-mining engineer plans mining operations. Coal-mining engineers determine whether coal deposits can be mined profitably. They design mine shafts and tunnels and equipment that cuts coal. They select explosives to blast coal deposits. They decide where to dig pits, where to put excavated rock and soil, and how to restore the land around mine sites.

This coal mining engineer is skilled at designing safe excavation sites.

An environmental officer works to protect or improve environmental quality and public health. Environmental officers work to control pollution and remedy environmental damages. They make sure that environmental laws and regulations are followed.

This environmental officer's test results will influence local mining policies.

Geoscenario Introduction: Yellowstone Hotspot

Yellowstone is one of America's most beloved national parks. Did you know that its unique scenery is the result of the area's geology?

Yellowstone National Park lies in a volcanic **caldera**, an area that collapsed after an eruption. Below the caldera is a **hotspot**. There, huge amounts of magma sit just below Earth's surface. In this geoscenario, you'll learn some of the geologic secrets that make Yellowstone such a special place.

Hydrothermal Features

Hot springs are naturally warm bodies of water. Hot magma heats water underground to near boiling. Some organisms still manage to live in these springs.

Its vivid colors and huge size make Grand Prismatic Spring the most photographed feature at Yellowstone. Extremely hot water rises 37 m from a crack in Earth's crust to form this hot spring.

The water in **mud pots** tends to be acidic. It dissolves the surrounding rock. Hot water mixes with the dissolved rock to create bubbly pots.

Other hydrothermal features include **fumaroles** and **geysers**. Fumaroles are cracks that allow steam to escape continuously. Geysers are hot springs that trap steam underground. As pressure builds, the steam erupts. Yellowstone has more than 300 geysers, more than any other place in the world. Yellowstone's most famous geyser is Old Faithful. While it is neither the largest nor the most regular geyser, it is certainly the most watched.

Hotspot Theory

Most earthquakes and volcanic eruptions occur near plate boundaries, but there are some exceptions. In 1963, John Tuzo Wilson (1908–1993) came up with a theory for these exceptions. He described stationary magma chambers beneath the crust. These hotspots can cause volcanic activity far from any **plate boundary**.

Recent technology has allowed scientists to create a picture of the magma chamber under the Yellowstone Hotspot. The chamber is huge. It sits underneath parts of Montana, Wyoming, and Idaho.

Under the Yellowstone Hotspot

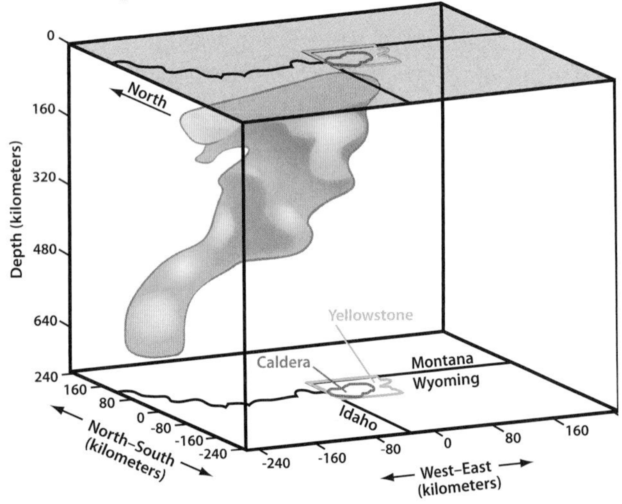

Seismologists at the University of Utah and the US Geological Survey mapped the location of this magma chamber, which sits under three US states and Yellowstone National Park. The caldera, or crater-like depression, from the last eruption (640,000 years ago) is outlined in red.

Unlike a hot spring, a geyser periodically shoots heated groundwater and steam up into the air. Yellowstone's Old Faithful geyser is world famous.

Path of Hawaiian Hotspot

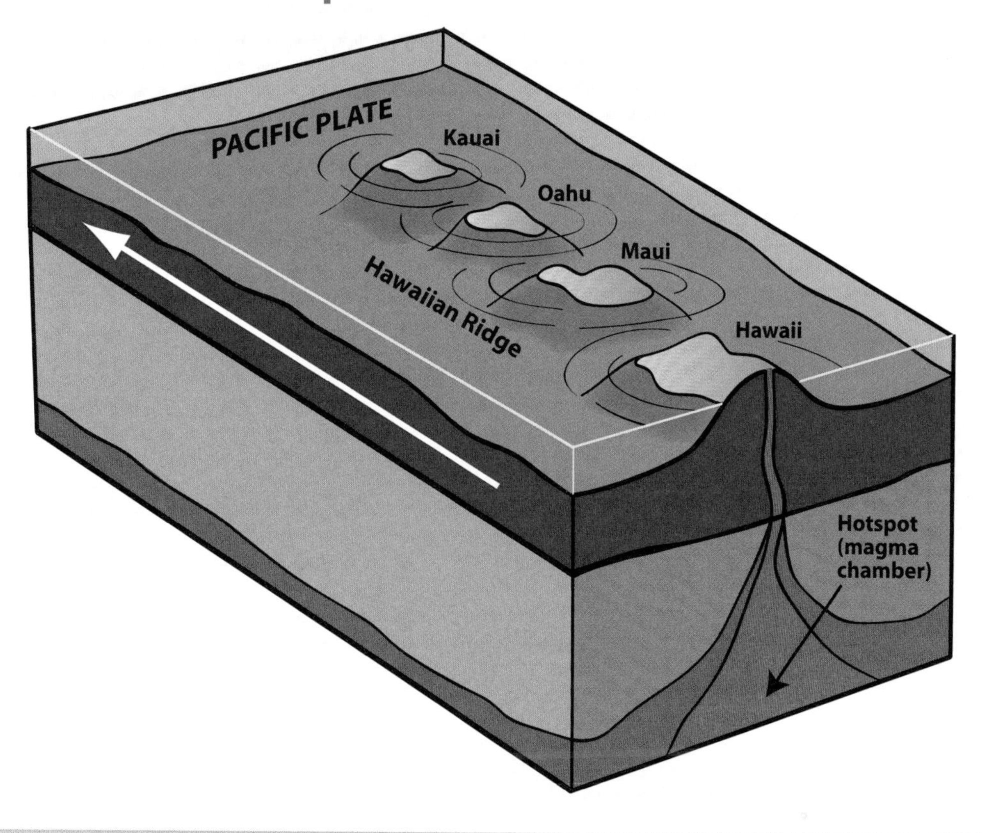

As the Pacific Plate has slowly moved north over a hotspot, the volcanic island chain of Hawaii has formed. Magma breaking through the crust at the hotspot is currently building the Big Island.

Understanding the Path of Hotspots

Hotspots appear to move over time. Evidence indicates, however, that what moves is a tectonic plate. Wilson suggested that the Pacific Plate moved northward over the fixed Hawaiian Hotspot. The island of Hawaii is the youngest and most volcanically active island. The other Hawaiian islands get older and less active as you move north.

Likewise, the North American Plate has been moving over the fixed Yellowstone Hotspot. The hotspot is melting a path through the Rocky Mountains.

Supereruptions

The past three eruptions from Yellowstone have been **supereruptions**. The Lava Creek eruption, 640,000 years ago, was one of the largest in history. It covered large areas with thick ash, as shown on the map on the next page.

Take a moment to consider the volume of some major eruptions. Mount St. Helens erupted in 1980 and ejected about 1 cubic kilometer (km^3) of volcanic material. How does that compare to Yellowstone? When the hotspot erupted 640,000 years ago, it ejected over 1,000 km^3 of volcanic materials. When it erupted 2.1 million years ago, it ejected an astonishing 25,000 km^3 of ash and debris. Ash covered much of what is now the United States.

The magma chamber below Yellowstone may cause another supereruption in the future. Geologists monitor the area carefully for signs of geologic activity. They hope to predict future eruptions and help keep people safe.

Geothermal energy use

Water heated underground that comes to the surface makes geothermal (heat from Earth) energy available to humans. Geothermal energy from geysers can be used to produce electricity. This is a **renewable** energy source, because it is constantly replenished. Should the geothermal energy in Yellowstone and the surrounding areas be used to produce electric power? Scientists and engineers are conducting studies to explore this. The studies consider the impact electric generation would have on the national park.

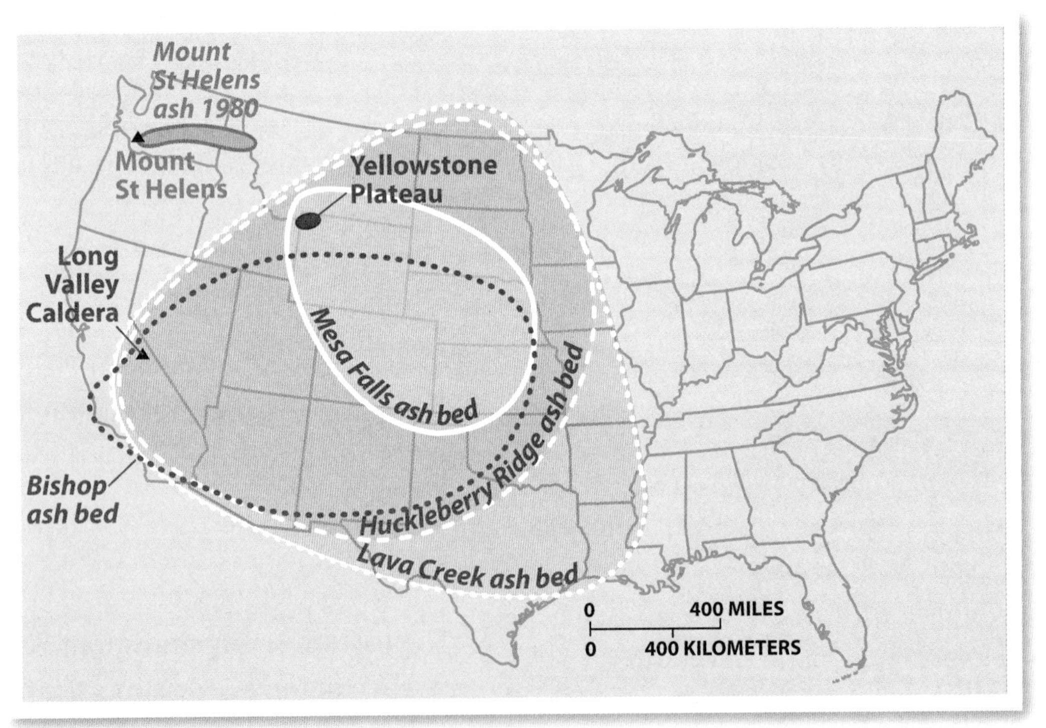

Geologists compare remnants of ash beds from ancient eruptions to ash beds of recent eruptions. They can infer the size of ancient, massive volcanic events from these comparisons.

Geoscenario Specialists: Yellowstone Hotspot

This page describes some of the careers chosen by people who might study hotspots. You will choose one of these careers for your research in the class project.

A geologist interprets and explains information about rocks, rock formations, and landforms. Geologists use patterns and evidence to understand the greater processes that shape Earth.

This geologist examines rock debris thrown from a geyser.

A seismologist studies earthquakes and the effects of waves in Earth's crust. Seismologists are especially interested in volcanoes and **tectonic** boundaries. They can create three-dimensional images of magma chambers that fuel hotspots.

This seismologist carries equipment to set up a remote sensing station.

A volcanologist studies the formation and behavior of volcanoes. Volcanologists aim to understand past eruptions and predict events. They often visit active volcanoes. They study old volcanic layers found within rock columns.

These volcanologists analyze data from recent volcanic activity.

A renewable energy consultant has a broad knowledge of energy issues for various industries. Renewable energy consultants help businesses improve their energy efficiency, energy consumption, and carbon footprint.

This renewable energy consultant attends an international conference on energy use.

Geoscenario Introduction: Oil

Oil is a part of every American's daily life. The cars we drive, the buses we ride, even the plastic that makes up your binder, pen, or comb all rely on oil.

Some people are eager to drill for **oil** wherever they find it. We now know that using oil can cause health and environmental concerns.

How will oil use change as Earth's population (and demand for energy) grows?

In this geoscenario, you'll learn about how oil formed, how people get oil, and some of the issues for oil use in the future.

Reservoirs of liquid petroleum trapped in underground rock are reached by drilling wells, typically more than 1.6 km deep. Then, pumps like these draw the so-called "black gold" to the surface.

The Geology of Oil

Oil is an excellent fuel because it is energy dense. A quantity of oil releases far more energy as it burns than the same mass of wood or coal. The key to its stored energy lies in its formation.

The story of oil starts millions of years ago, with ancient organisms. Energy from photosynthesis (plants) or from eating (animals) is stored in molecules in the organisms' bodies. Millions of years ago, areas that are now sedimentary rock, such as sandstone, limestone, and shale, were underwater in basins. As layers of sediment built up, dead organisms became trapped in the layers. Leaves and algae collected in muddy swamps. Algae and *foraminifera* (microorganisms) mixed with the sandy bottom of the ocean. Sediment trapped countless microscopic organisms and the remains of other organisms.

Heat and pressure turned the sediment into rock. They also transformed the remains of the organisms into oil over millions of years. Burning oil releases the energy stored in molecules by those ancient organisms.

Oil in the Gulf of Mexico

Oil, along with coal and natural gas, is a **fossil fuel**. When used, fossil fuels release carbon dioxide into the atmosphere. Carbon dioxide is a greenhouse gas.

Fossil fuels are called nonrenewable resources because they develop over millions of years. When we use up the supplies stored below Earth's surface, we cannot make more. So people are willing to do risky things to capture oil.

The 2010 oil spill in the Gulf of Mexico began with an explosion on the *Deepwater Horizon* oil rig on April 20. After much difficulty, the leak was permanently sealed on September 19. Approximately 4 million barrels of **crude oil** leaked into the Gulf waters. (One barrel equals about 160 liters [L].) The oil spill created an ecological disaster for plants and animals living in the Gulf and along the shoreline.

After the disaster, all drilling in the Gulf stopped. Governmental boards reviewed safety procedures and policies. Then, the US government allowed companies to resume drilling and exploring in deep ocean water.

Offshore drilling comes with many risks. Oil companies have been drilling deeper and deeper to meet the demand for petroleum products. Today, the Gulf of Mexico has nearly 4,000 active oil rigs.

Oil Products

When oil is first removed from the ground, it is called crude oil, or petroleum. Crude oil cannot go directly into an automobile or airplane. It is refined to break it into different substances for different products.

Tides and currents carry spilled oil to shorelines, where it coats the land, clogs the feathers and fur of marine animals, and smothers plants. Ecological recovery is long and costly.

Oil spills are most often caused by accidents involving pipelines, tankers and barges, or offshore drilling platforms, like this rig off the southern coast of California.

Products Made from a Barrel of Crude Oil (Gallons) (2009)

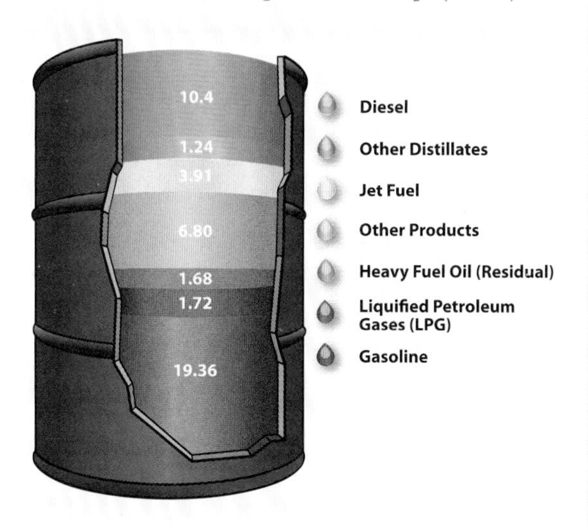

Once extracted from the ground, crude oil is shipped to refineries. There, different components are separated and processed for different uses.

Most petroleum products are fuels, such as gasoline and diesel. Burning the fuels heats buildings and powers engines. Oil is also used to manufacture plastics and chemicals, such as drugs, fertilizer, and pesticides.

You may have noticed that the price of gasoline changes over time. The price of a barrel of oil depends on supply and demand. Oil prices affect gas prices at the pump. The United States does not produce enough crude oil to power all the vehicles on the road.

Oil Production and Consumption

Many countries around the world produce oil, which means they pump the oil from the ground. Remember that people do not actually make oil. Oil production occurs naturally over millions of years.

The United States consumes more oil than it produces. More than half the oil it uses is imported from other countries (Canada, Saudi Arabia, Venezuela, Mexico, and Colombia). About one-third comes from the Gulf of Mexico.

More than 90 million barrels of oil are used every day throughout the world. The United States uses over 20 percent of this total. However, Americans make up less than 5 percent of the world's population.

Go to FOSSweb.com to review updated oil production and consumption data

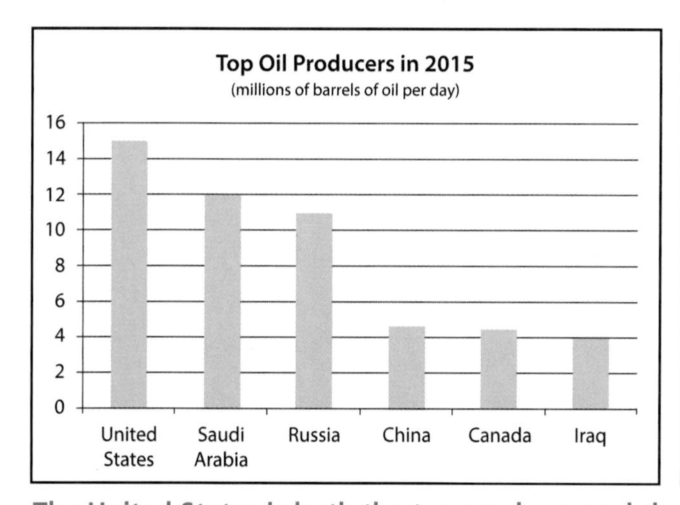

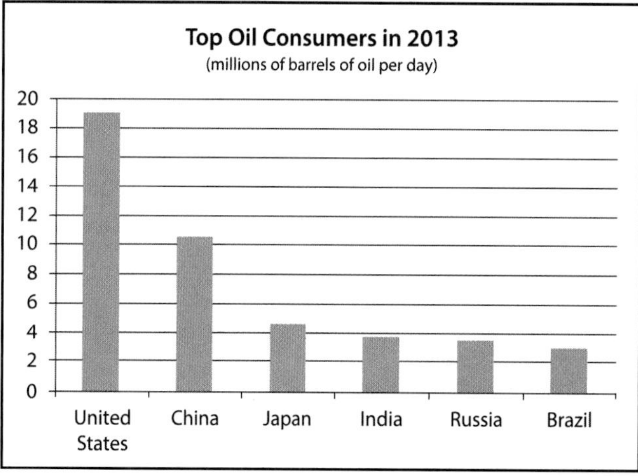

The United States is both the top producer and the top consumer of oil. Many people are looking for alternative, renewable sources of energy to meet our needs.

Geoscenario Specialists: Oil

This page describes some of the careers chosen by people who might study oil. You will choose one of these careers for your research in the class project.

A petroleum geologist identifies where oil can be found. Petroleum geologists analyze geologic maps and look for signs of oil out in the field. Oil companies depend on petroleum geologists to tell them where to drill for oil.

This petroleum geologist visits an oil drilling site on the coast.

Petroleum engineers apply their understanding of chemistry to transform crude oil into a variety of products. Then, they design the equipment to do it. Some of these products include plastics, medicine, pesticides, fertilizers, paint, and makeup.

This petroleum engineer examines oil samples in a chemistry lab.

An environmental scientist needs to understand several various fields of science. Environmental scientists study how to minimize our impact on the environment. They work for the government as well as private companies to make sure that they follow environmental policies.

These environmental scientists check for oil contamination on a marsh island.

An energy-policy analyst advises the government and other organizations on trends in energy. Energy-policy analysts inform lawmakers about how to address problems with nonrenewable resources.

These energy policy analysts present findings that may decrease fossil fuel use.

Research Careers in the Lab and Field

Antarctica's Mount Erebus has been an active volcano for 1.3 million years. It has frequent ash eruptions and a rare lava lake that continually bubbles in its inner crater.

How do we know so much about volcanoes and earthquakes? How can we learn more about these catastrophic events? Read about two young scientists who are doing research in this field.

Dr. Kayla Iacovino: Volcanologist

Dr. Kayla Iacovino is a **volcanologist**. Her work in the lab includes heating rocks and minerals to over 1,000 degrees Celsius (°C) to simulate rock formation deep in Earth. In the field, Dr. Iacovino travels to volcanoes to collect data about rock composition and volcanic activity.

FOSS: When did you start to explore geology?

KI: As a kid, I liked to collect rocks. In college, I realized that geology is the study of Earth, the study of the stuff that's around us, the stuff that makes up the universe. It's the ultimate application of biology, physics, chemistry, and math.

I got interested in geology as a career when a researcher put out a job advertisement that said, "Get paid to melt rocks!" I thought, "That's awesome! Of course I want to go do that." I decided that I want to study how Earth works, particularly volcanoes.

FOSS: What did you work on in that lab?

KI: I focused on the origins of rocks. The igneous rocks that we see on the surface have come from somewhere deep in Earth. What have the rocks gone through to get here? There are clues in the rock that we can look for. Geologists can try to recreate the rocks we see on the surface to understand that.

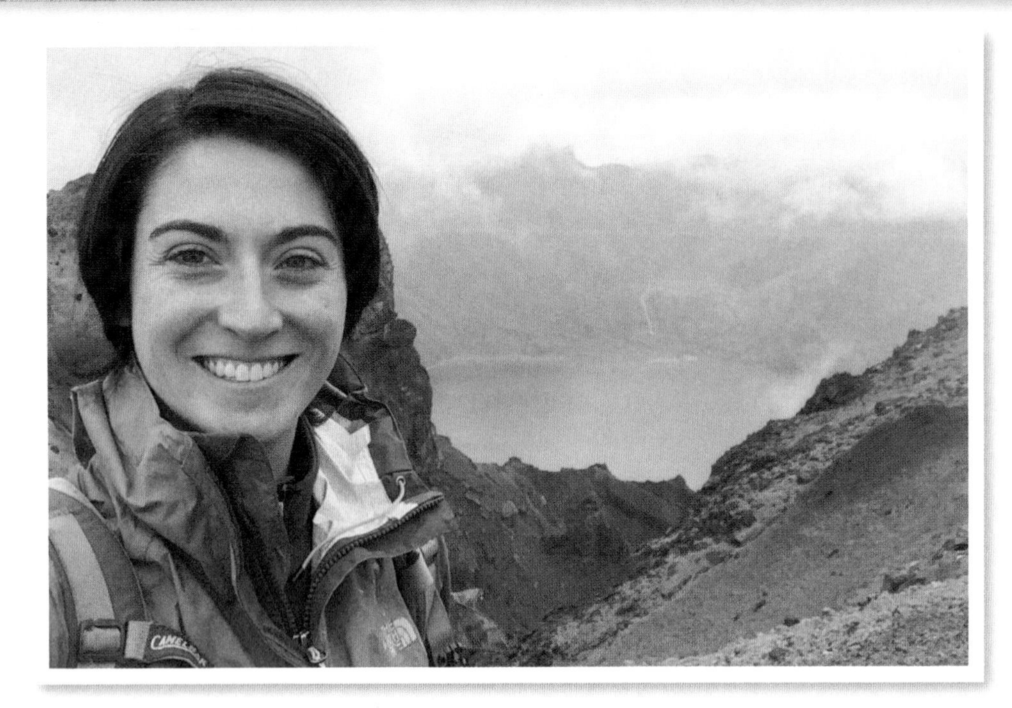

Dr. Kayla Iacovino combines fieldwork, lab experiments, and modeling to investigate Earth's rock cycle.

In the lab, I can take a material, cook it, and see how the material changes when I apply different pressures and temperatures, or change its chemistry. If I can reproduce something like natural rocks, that tells me that I've figured out the precise temperature, pressure, and chemical conditions necessary to create that rock. So I can say where it came from. The rock tells you a story about all the things that have happened to it along its life cycle. The geologist has the tools and the knowledge to decode that record.

FOSS: When you are not in your lab, what kind of research do you do in the field?

KI: At the end of my PhD research, I studied a volcano called Mount Paektu. It is half in China and half in North Korea. We were collaborating with North Korean geologists in the country.

Mount Paektu has been difficult to study, but it has a lot of stories to tell. We were able to ask very fundamental questions about the volcano. Why is the volcano here? It's not at a plate boundary. What is happening hundreds of kilometers deep in the Earth? What is causing magma to form? How did one of its ancient eruptions affect the global climate?

The most awe-inspiring volcano I've ever been to is Mt. Erebus in Antarctica. It's famous because it has a lava lake. You can stand at the rim of the crater and look down to see inside the volcano. There's a pit of lava that's boiling and churning away, and you can see inside the volcano. Usually we see a sort of a dome of rocks and it's steaming, but here you can actually see the lava sitting there in the crater. That was incredible.

FOSS: Tell us a little bit about your current research.

KI: I'm studying subduction. Subduction is kind of like a conveyor belt that is bringing material from the surface down into the mantle. That causes the mantle to melt and form magma, which erupts from volcanoes. It's the rock cycle: bringing stuff down, bringing stuff back up. Subduction helps the materials in the Earth's interior and exterior mix.

But we can't go deep underground to see this mantle melting happen! I can try to recreate the subduction-zone process in the lab to understand how it happens.

I'm studying how this part of the rock cycle has changed Earth's atmosphere over the last several billion years. We think that when the processes of plate tectonics and subduction began on Earth, they may have changed the composition of the atmosphere. That could have affected the development of life on Earth.

FOSS: How does your field use new technology?

KI: Satellites record temperatures of lava and track gas and ash flows traveling across the surface of Earth. We use drones or quadcopters to collect data right around the volcano. They can fly around, take pictures, collect rocks, and even dunk tools into lava to take samples.

Dr. Iacovino heats silica glass in the lab to learn about properties of earth materials.

Studying volcanoes and earth materials can reveal information about processes under Earth's surface.

FOSS: Can volcanologists predict eruptions?

KI: If anyone tells you they can predict a volcanic eruption, they're either lying or they're wrong. But we're getting better at it. We may see a lot of small earthquakes happening at a volcano, and the temperature and composition of the volcanic gases changing. These things can indicate that maybe an eruption is going to happen. The reason it's so difficult is that every volcano is different. You might be able to understand the signals of one volcano very well, like when the sulfur levels go up, it's going to erupt. But you could take that knowledge to another volcano and have it be untrue. We need to know what's going on deep down below the surface, in the belly of the volcano, that's causing these changes. Volcanology is a relatively young science, especially compared to things like physics and chemistry. We definitely don't know everything about volcanoes. We have a lot to learn!

Dr. Angela Chung: Geophysicist

Dr. Angela Chung is a geophysicist who develops computer programs with the ShakeAlert project at the University of California, Berkeley. ShakeAlert collects earthquake data from sensors in earthquake zones. Computers analyze the data and warn people who live near the epicenter that an earthquake is coming. People may get only a few second's warning, but that's enough time to move to safety.

Dr. Angela Chung works to improve the speed and reliability of ShakeAlert's early warning system. The goal is to rapidly detect potentially damaging earthquakes and issue public alerts.

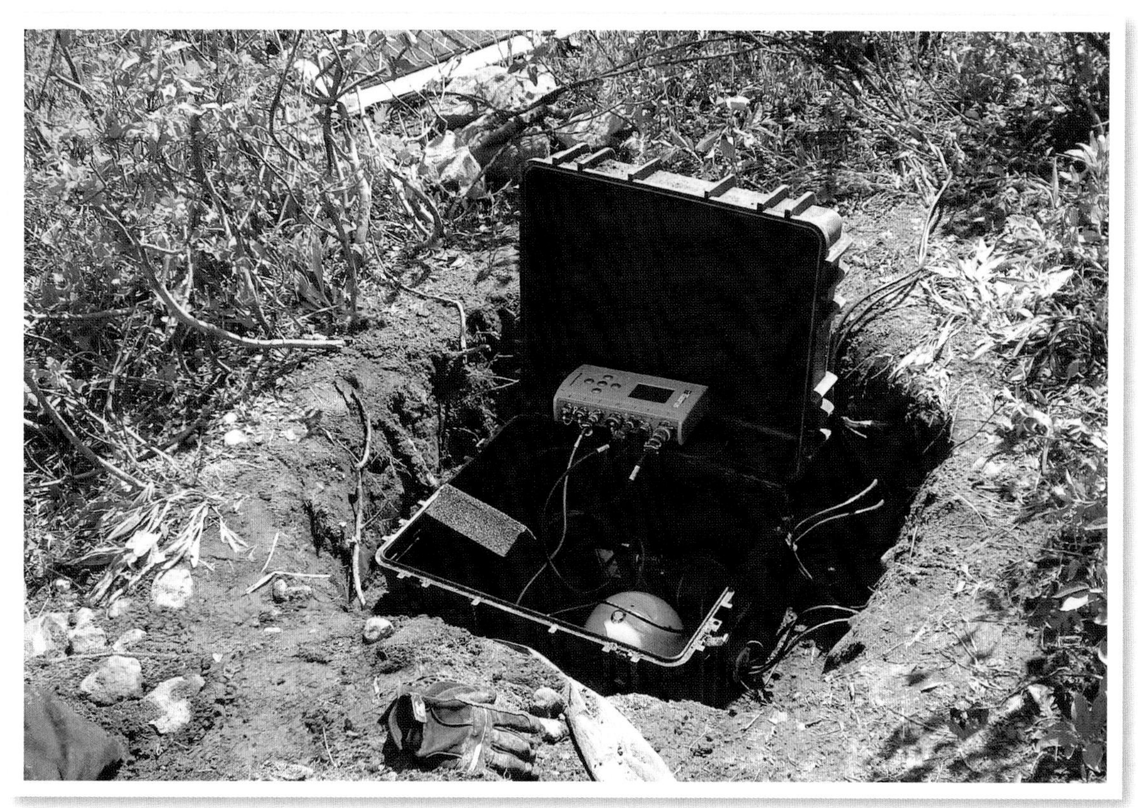

Dr. Chung installed this earthquake sensor in Yellowstone National Park, a seismically active area.

FOSS: How did you become interested in geology and physics?

AC: My dad is an electrical engineer and a musician. He did his own home recording. I found it fascinating. When you are doing a home recording, you can watch the signals, the wave forms, that you're creating. That was one of the things that sparked my curiosity about signals and how they vary. Now I look at wave-form signals from earthquakes instead of music. I started as a physics major at UC Berkeley, but I took an earth science class and found that I really liked it. So I switched to geophysics, which combines the two fields.

FOSS: What is your role in the ShakeAlert project?

AC: I help improve our earthquake detections, to make them as good and as fast as possible. We can tell there's an earthquake only when we've recorded it at a station. I improve our computer analysis of these data. We sometimes get signals that are not from earthquakes. There might be construction going on near the station, or the station has been damaged somehow. I write computer programs to filter out those bad signals.

FOSS: Can geologists predict earthquakes?

AC: Nobody can predict them at this point. But we can say where there is likely to be an earthquake. ShakeAlert doesn't predict earthquakes. It monitors them as they occur and estimates the ground shaking in surrounding areas.

FOSS: How does ShakeAlert use new technology?

AC: Stations are constantly streaming high-speed data. When an earthquake occurs, we find out within a few seconds and can tell you where it is and how big it is. This speed wasn't possible before the Internet. We are trying to develop ways to use sensors in cell phones. That way we can collect a lot more data. But cell phones are always moving around with people. We need to develop better computer programs to determine whether it's an earthquake or just someone walking.

FOSS: What is daily life like at your work?

AC: I spend a lot of time at the computer. We are developing some new, low-cost sensors. Once we have some prototypes, I'll be able to go out and install some of those. That's fun. The best way to test a network of sensors is where there are lots of earthquakes. That happens immediately following a big earthquake because there are lots and lots of aftershocks. I went to Chile after their 8.8 magnitude earthquake in 2010 and set up sensors all around the country. I then went to New Zealand in 2011 after their 7.1 magnitude earthquake.

FOSS: What are the next steps for ShakeAlert?

Dr. Chung visited Chile after a magnitude 8.8 earthquake in 2010. She set up and tested sensors during the aftershocks.

AC: We want to have public alerts up and running in the next few years. Before then, we have to make sure that the system works really well. If we're telling thousands of people that an earthquake is heading their way, we need to be really sure. We also want to improve our detection in areas where there aren't many sensors. If we have more stations, we can more easily pinpoint exactly where an earthquake is occurring.

We also do public outreach. Earthquakes are big scary things, but when you learn about them, you can prepare for them.

Take Note

What questions in geology might you want to help find the answer to?

Images and Data

Images and Data Table of Contents

Getting to Know the Grand Canyon

Grand Canyon Vital Statistics	
Length from Lees Ferry to Grand Wash Cliffs	443 kilometers (275 miles)
Elevation of the river at Lees Ferry	947 meters (3,107 feet)
Drop in river elevation between Lees Ferry and Grand Wash Cliffs	575 meters (1,886 feet)
Average width from rim to rim	16 kilometers (10 miles)
Greatest width	29 kilometers (18 miles)
Smallest width	8 kilometers (5 miles)
Depth from North Rim	1,737 meters (5,700 feet)
Average elevation of North Rim	2,438 meters (8,000 feet)
North Rim highest elevation (Point Imperial)	2,682 meters (8,800 feet)
Average elevation of South Rim	2,072 meters (6,800 feet)
South Rim highest elevation (Grandview Point)	2,255 meters (7,400 feet)
Canyon depth near Grand Canyon Village	1,524 meters (5,000 feet)
Colorado River water temperature before 1963	26.6°C (80°F)
Colorado River water temperature after 1963	5.6°C (42°F)

The Grand Canyon and the Colorado River

- The Colorado River was first named Rio Colorado, or Red River, by the Spanish.
- The river's color is now blue green. When Glen Canyon Dam was completed in 1963, it trapped the sediments that made the river reddish brown.
- The Colorado River still turns red when flash floods carry sediments into it from side canyons.
- The Colorado River is the main river in the American Southwest. It drains parts of Wyoming, Colorado, Utah, New Mexico, Arizona, Nevada, and California.
- The headwaters of the Colorado River are in Rocky Mountain National Park in Colorado. The river starts at an elevation of 2,746 meters (m). It flows southwest toward the Gulf of California and the Pacific Ocean.
- Some native fish that once lived in the river have become extinct. Others are endangered. The fish populations shrank because damming changed the water temperature.

People at the Grand Canyon

- Native Americans have been living in the Grand Canyon area for at least 4,000 years.
- Twig figures made of split willow provide evidence that these early settlers lived in the Nankoweap Canyon area.
- In 1540, García López de Cárdenas (dates unknown) was the first European to view the Grand Canyon. Francisco Vásquez de Coronado (1510–1554) sent Cárdenas north from Mexico to search for the fabled seven cities of gold.
- Cárdenas and his party spent 3 days at the canyon, trying to get down to the river. They ran out of supplies and had to turn back.
- In 1869, John Wesley Powell (1834–1902) became the first person to document the exploration of the entire length of the canyon.
- President Theodore Roosevelt (1858–1919) worked to make the Grand Canyon area a national park. It took 11 years to become official, in 1919.

Visiting the Grand Canyon

- It takes most people at least 1 full day to hike to the bottom of the Grand Canyon. It takes 2 full days to hike back safely.
- A motor-powered raft can travel along the river from Lees Ferry to Grand Wash Cliffs in about a week. It takes 2 or 3 weeks in an oar-powered raft or boat.
- If you could fly from Grand Canyon Village on the South Rim directly to the North Rim, you would travel only 19.3 kilometers (km) in just a few minutes.
- If you drove around the canyon from one rim to the other, you would travel 346 km, and it would take 5 hours.

Landforms Gallery

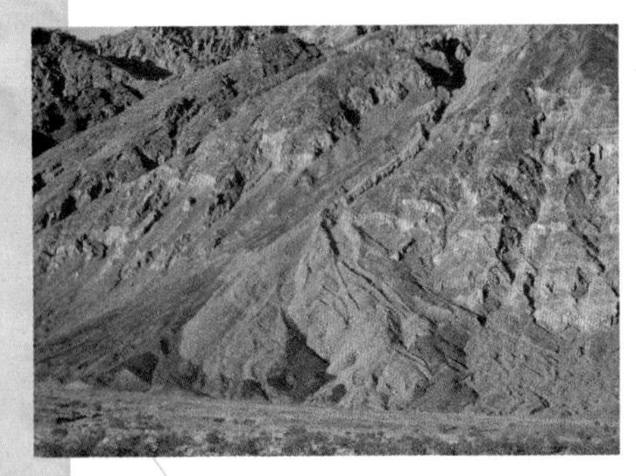

Alluvial fan: a fan-shaped deposit of rocks formed where a stream flows from a steep slope onto flatter land

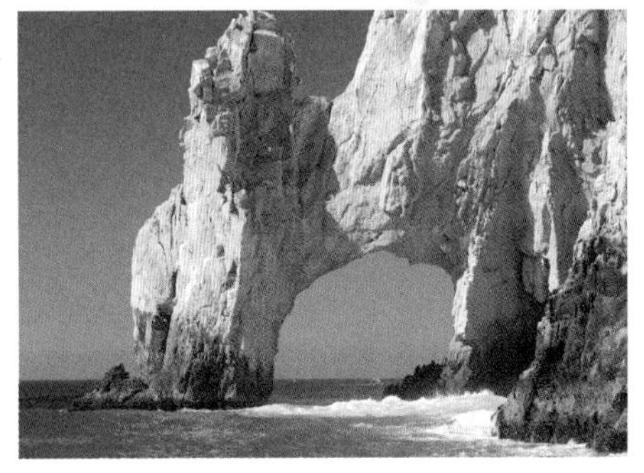

Arch: a curved rock bridge that forms when chemical and physical weathering weakens the center, and the rock erodes. Arches can form on land or near the coast, where waves batter and erode the centers.

Beach: a gently sloping shore next to a body of water washed by waves or tides, often covered by sand and pebbles

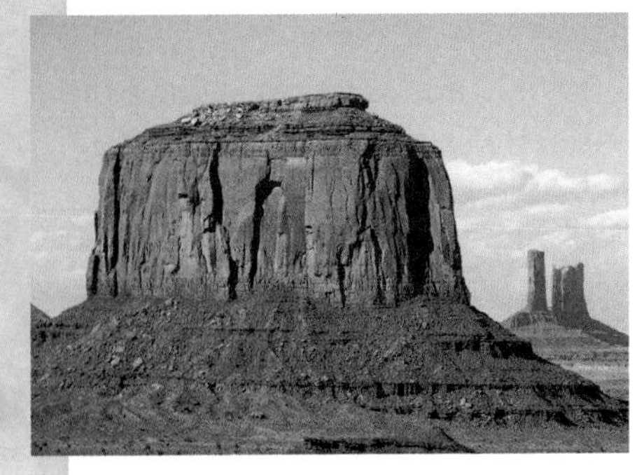

Butte: a hill with steep sides and a small flat top. A butte is smaller than a mesa.

Caldera: a hole that forms when the top of a volcano blows off or when the magma below the volcano drains away

Canyon: a V-shaped valley with steep sides eroded by a stream

Cinder cone: a volcano formed from a pile of cinders and other volcanic material blown out in an explosive eruption

Cliff: a high, steep face of rock

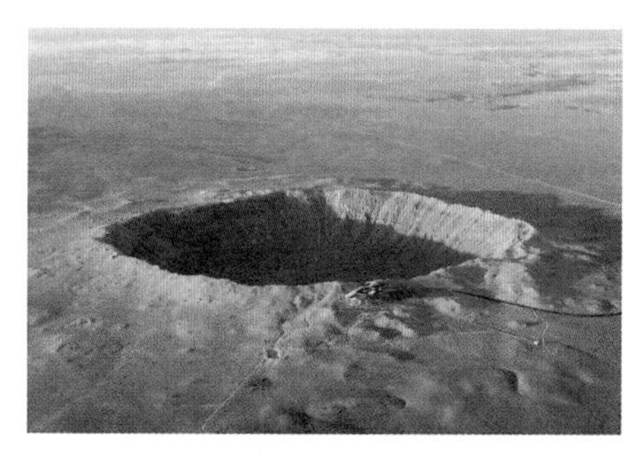

Crater: a bowl-shaped depression made by an explosion at the mouth of a volcano or geyser or by the impact of a falling object like a meteorite

Delta: an often fan-shaped deposit of earth materials at the mouth of a stream or river

Dome: a round shape formed when rocks like granite peel away at Earth's surface

Fault: a break in Earth's surface along which movement has occurred

Floodplain: land covered by water during a flood. Small particles, like sand and silt, are deposited on a floodplain.

Glacier: a large mass of ice moving slowly over land

Hoodoo: a formation where weak rocks erode away, leaving oddly shaped resistant rocks

Island: a piece of land surrounded by water

Lake: a medium-to-large body of water surrounded by land

Landslide: the pile of earth materials that results from their rapid movement down a slope

Levee: a stream bank that may stop land from flooding; can be natural or constructed

Meander: a curve or loop in a river or stream

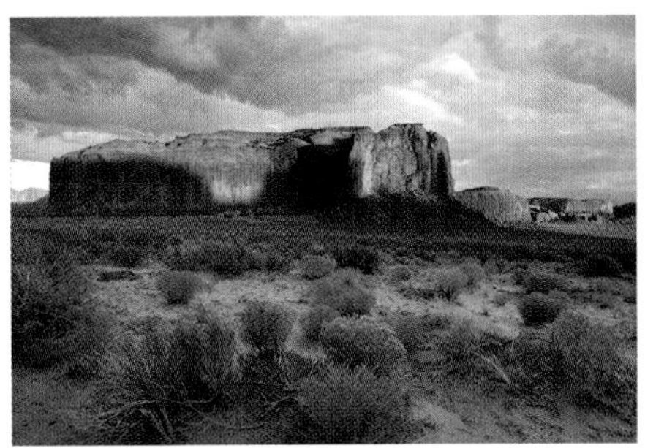

Mesa: a single, wide, flat-topped hill having at least one steep side

Moraine: a mound or ridge of unsorted rocks and soil carried and deposited by a glacier

Mountain: a large landform that stretches above the surrounding land in a limited area, usually with a peak

Pinnacle: a high, narrow piece of rock

Plain: a low area of Earth's surface that is often formed of flat-lying sediments

Plateau: a nearly level uplifted area composed of horizontal layers of rock

River: a large natural stream of water flowing in a channel

Sandbar: a long ridge of sand built up by river currents or ocean waves in shallow water

Sand dune: a ridge, mound, or hill of sand deposited by wind

Valley: a low area between hills or mountains where a stream or glacier flows. Stream valleys are V-shaped. Glacier valleys are often U-shaped.

Volcano: a mountain made of lava, cinders, and ash that have poured out through an opening in Earth's crust

Waterfall: a cascade of water falling from a height, formed when a river or stream flows over a cliff or steep incline

Landforms Vocabulary

Grand Canyon Map

Grand Canyon Views

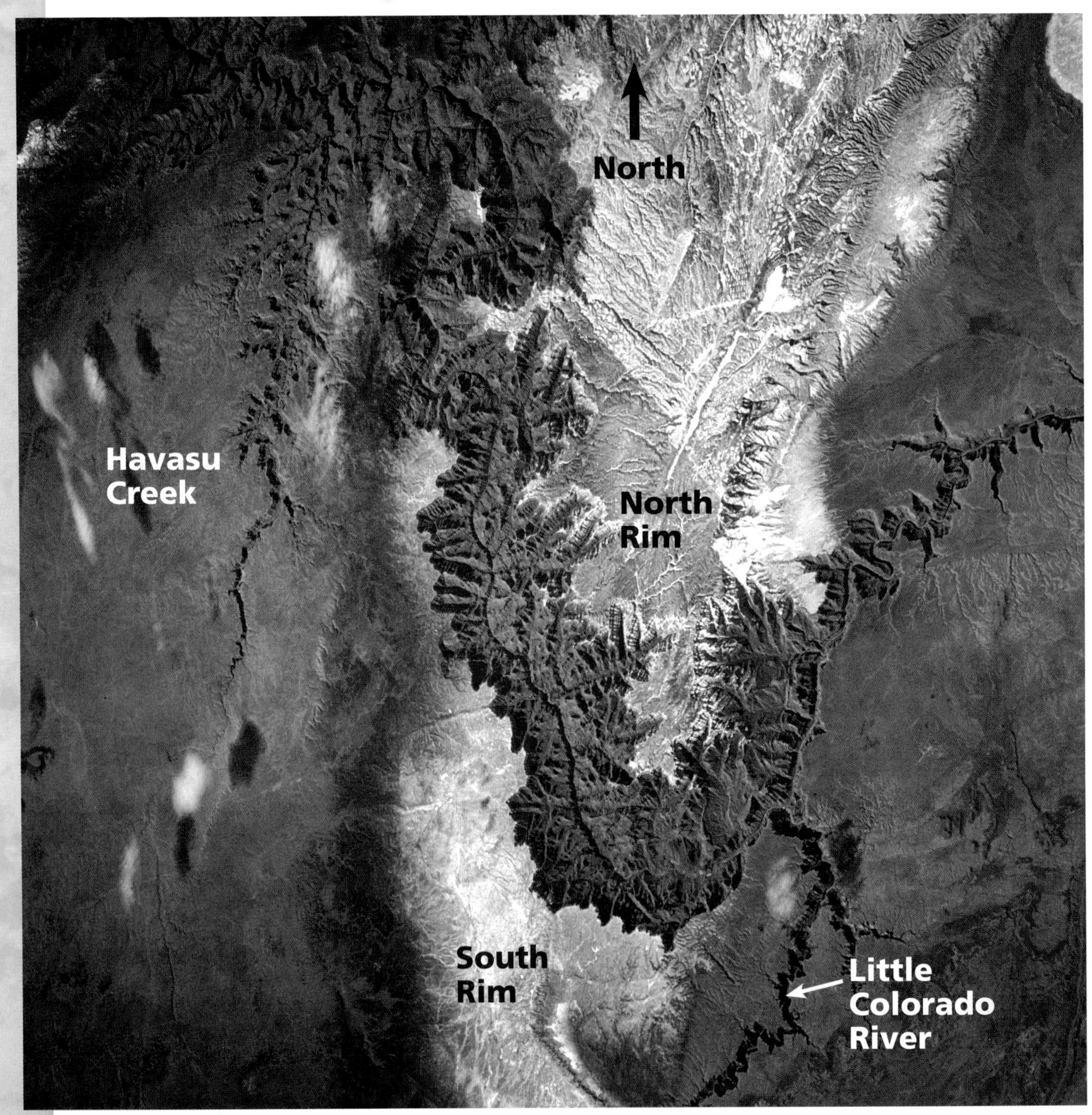

A space-shuttle view of the Grand Canyon

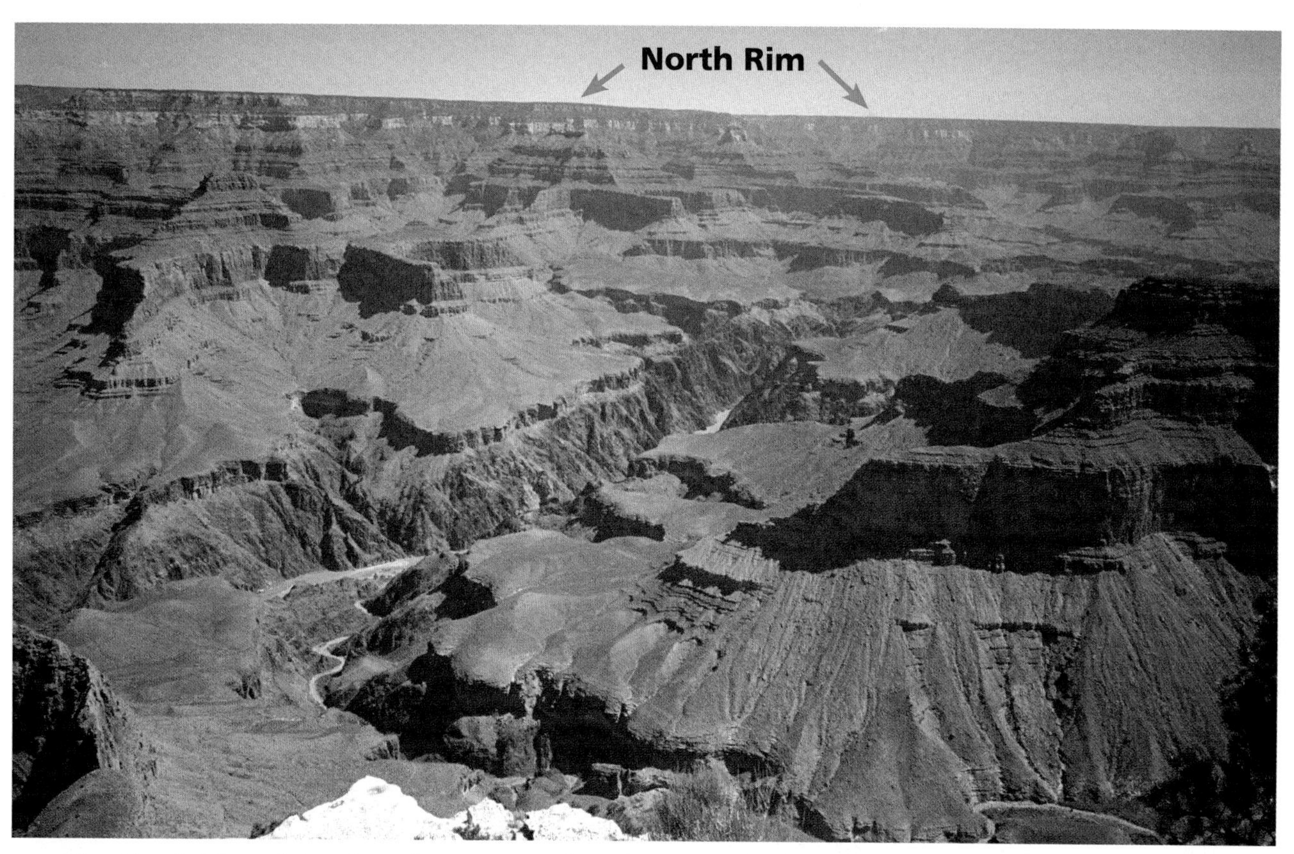

The North Rim of the Grand Canyon, taken from South Rim

River level in the Grand Canyon

South Kaibab Trail in Grand Canyon National Park

Stormy sky over the Grand Canyon

Granite Gorge rapids, taken from Plateau Point

The North Rim of the Grand Canyon, from South Kaibab Trail, South Rim

Human History in the Grand Canyon

8000 BCE	Paleo-Indians live at the Grand Canyon.
1100 CE	Prehistoric Pueblo peoples (Anasazi) inhabit the canyon.
1540	First Europeans visit canyon with Hopi guides.
1869	John Wesley Powell leads first expedition down the Colorado River.
1871	First photographs of the Grand Canyon from the rim are taken by Timothy O'Sullivan.
1893	President Benjamin Harrison creates Grand Canyon Forest Reserve.
1901	Train service to the Grand Canyon begins.
1902	The first automobile arrives at the Grand Canyon. Ellsworth and Emery Kolb establish their photography business at the canyon.
1903	President Theodore Roosevelt visits the Grand Canyon.
1905	The El Tovar Hotel opens for business.
1908	President Theodore Roosevelt creates Grand Canyon National Monument.
1915	Annual visitation to the canyon reaches 106,000.
1918	The US Congress creates Grand Canyon National Park.
1919	President Woodrow Wilson signs the Grand Canyon Park bill. Annual visitation to the park drops to 44,000 because of World War I.
1920	The park is officially dedicated.
1928	The Kaibab suspension bridge is constructed across the Colorado River.
1956	Construction of Glen Canyon Dam is authorized.
1963	Glen Canyon Dam is completed.
1969	Annual visitation exceeds 2,000,000.
1976	Annual visitation exceeds 3,000,000.
1991	National Park Service begins drafting a new general management plan.
1997	Annual visitation exceeds 5,000,000.
2007	The Hualapai Tribe opens a transparent skywalk over a section of the canyon.

The Powell Expedition Map

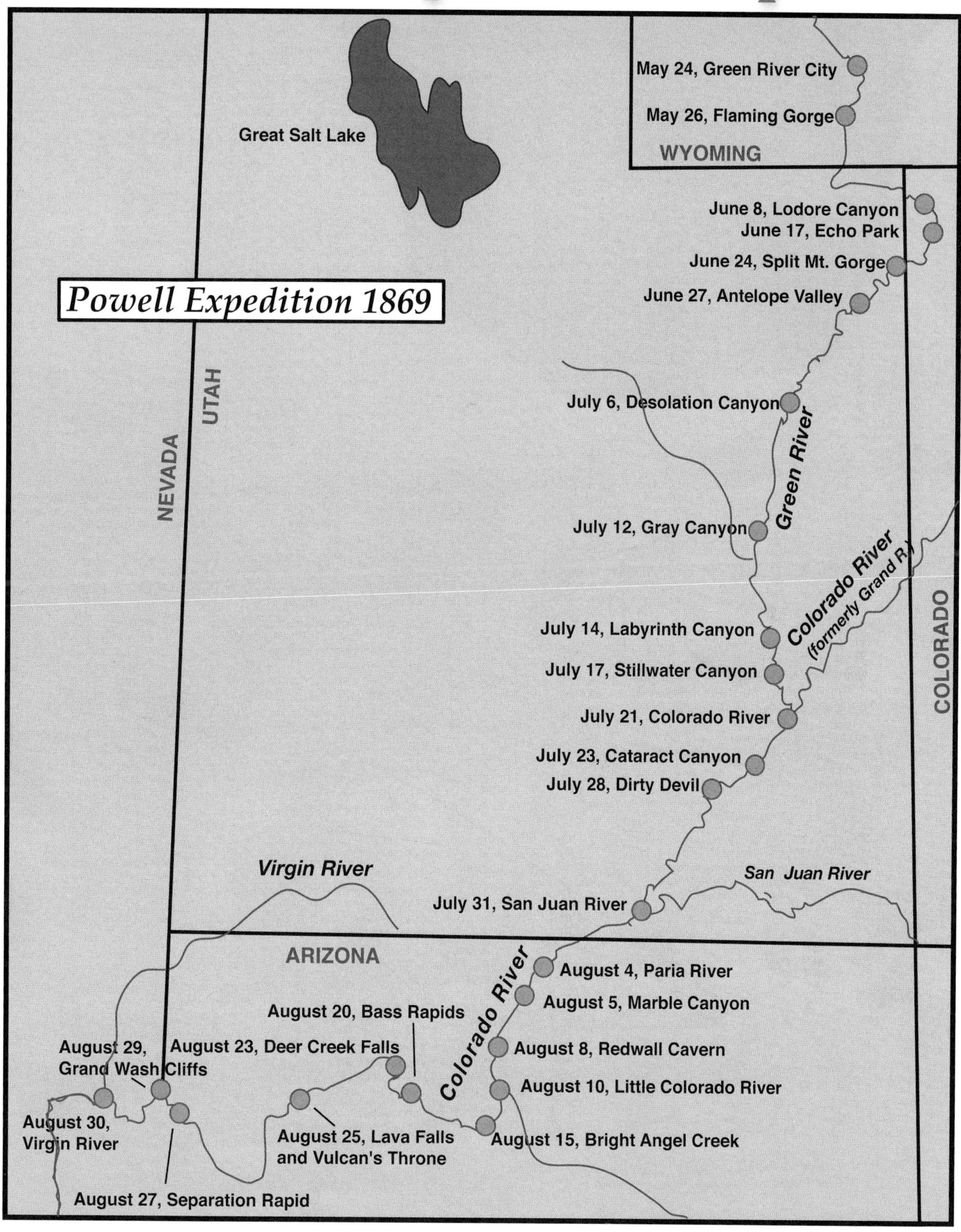

Mile 20

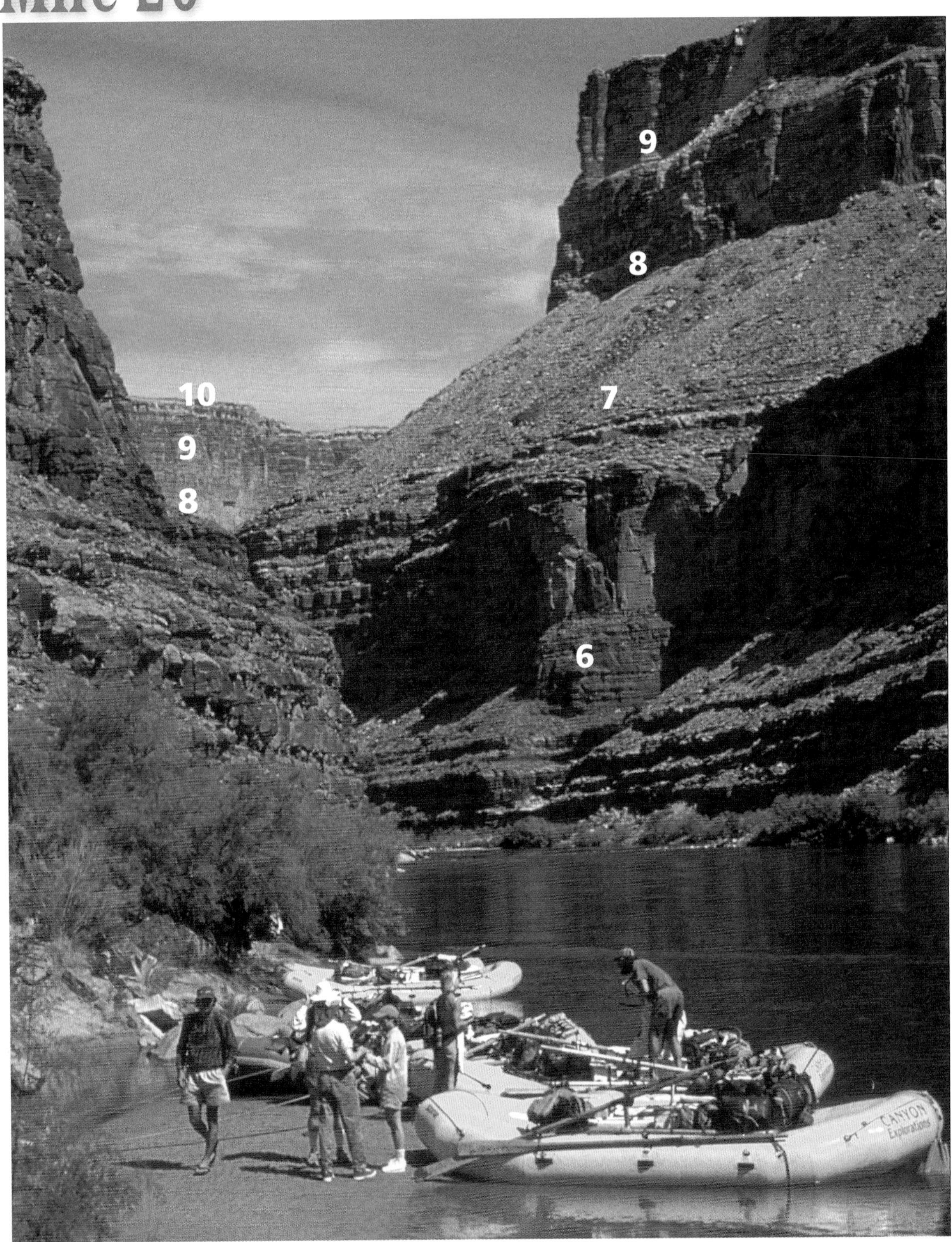

River elevation about 891 meters

Mile 52

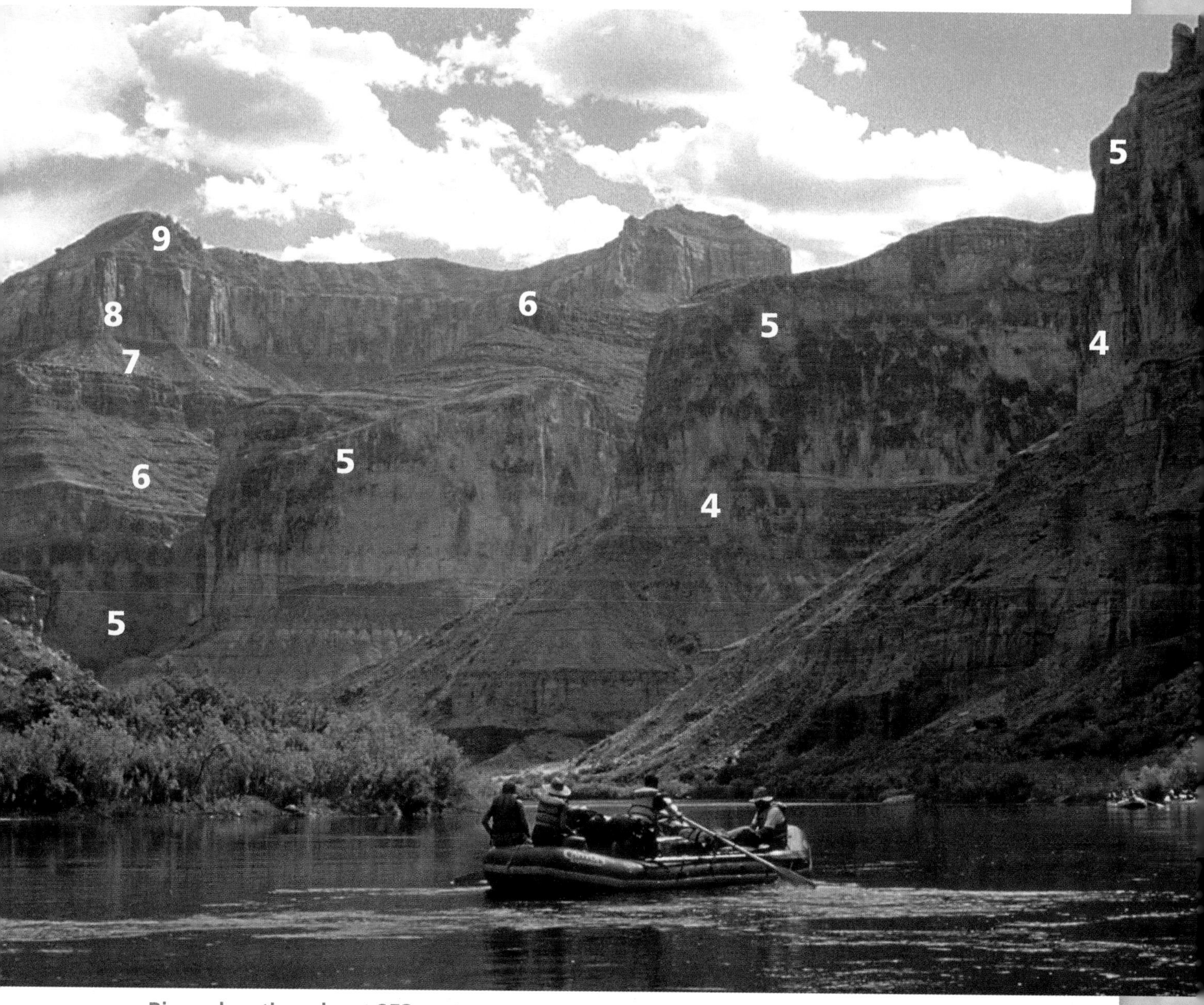

River elevation about 853 meters

Grand Canyon Rocks

Rock 10 from the rim

Rock 10 close-up

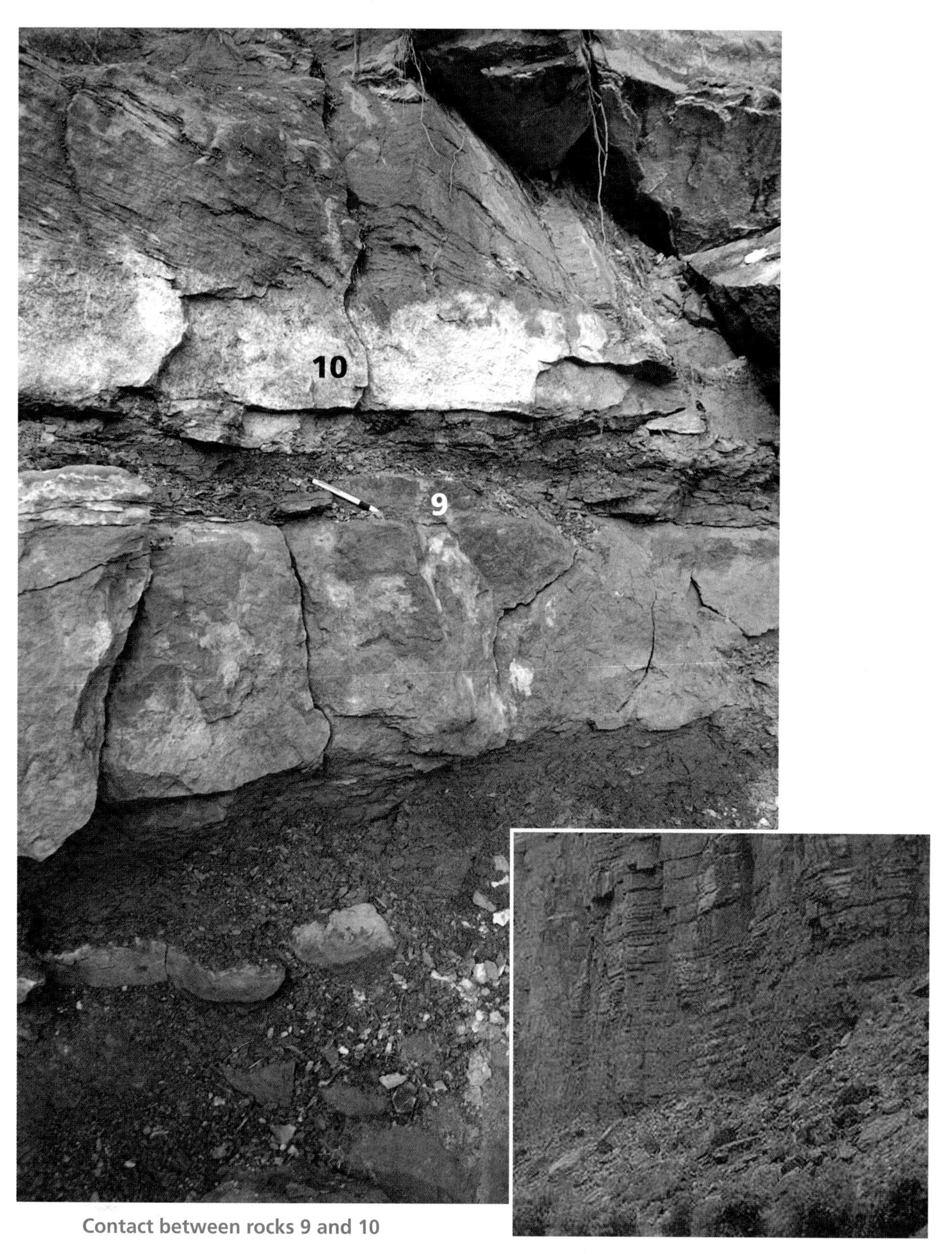

Contact between rocks 9 and 10

Rock 9, medium distance

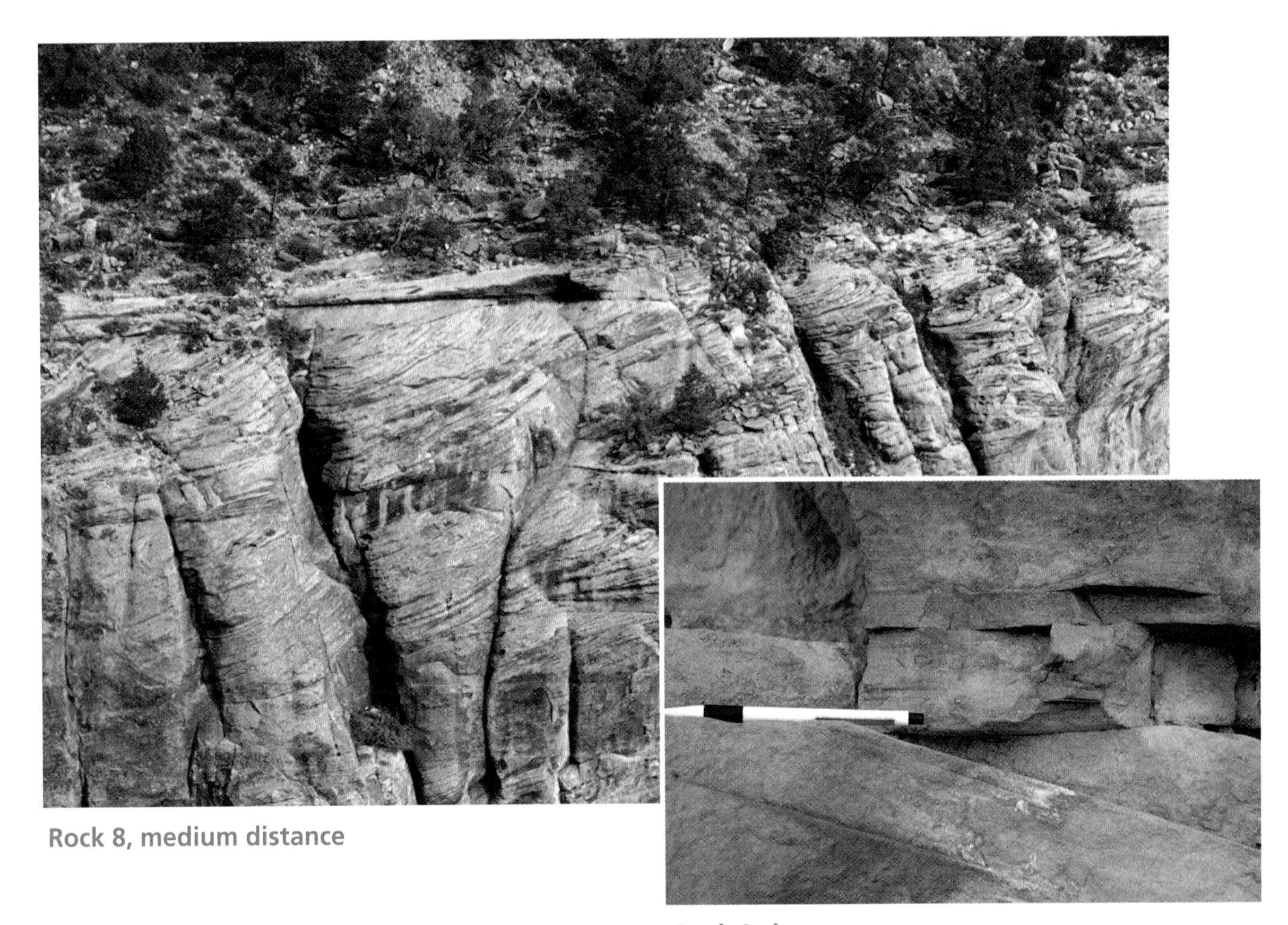

Rock 8, medium distance

Rock 8 close-up

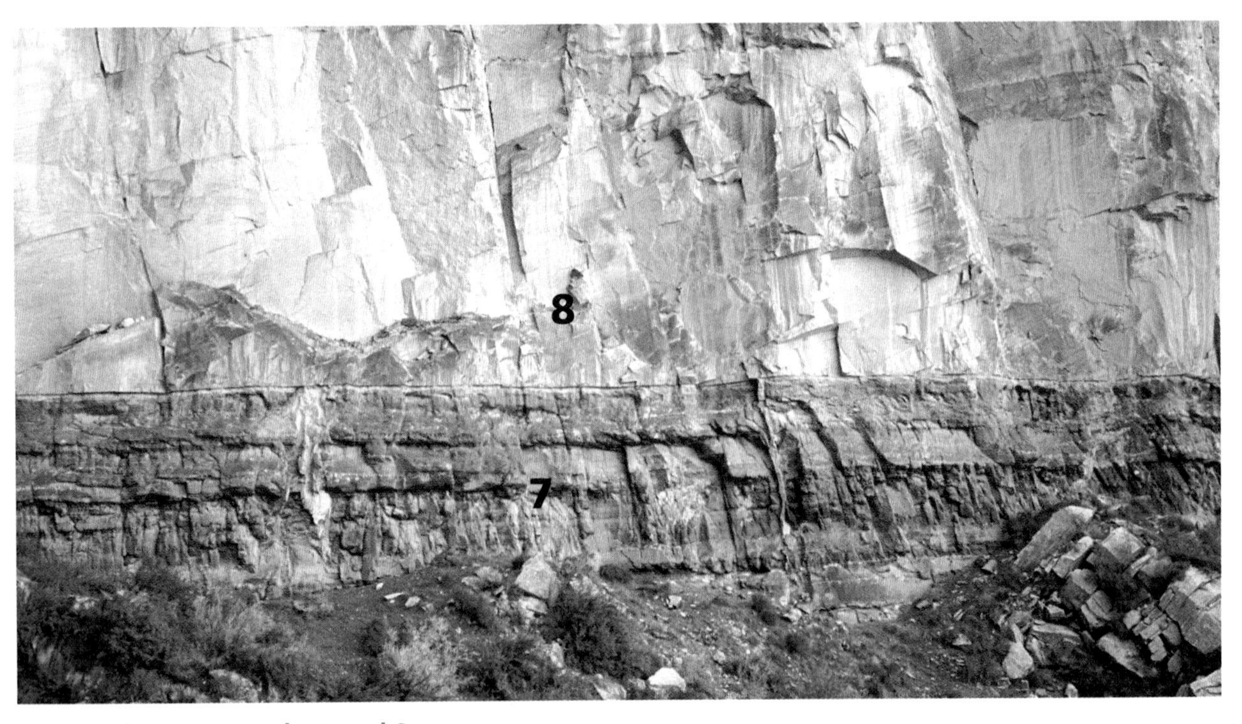

Contact between rocks 7 and 8

Rock 7 close-up

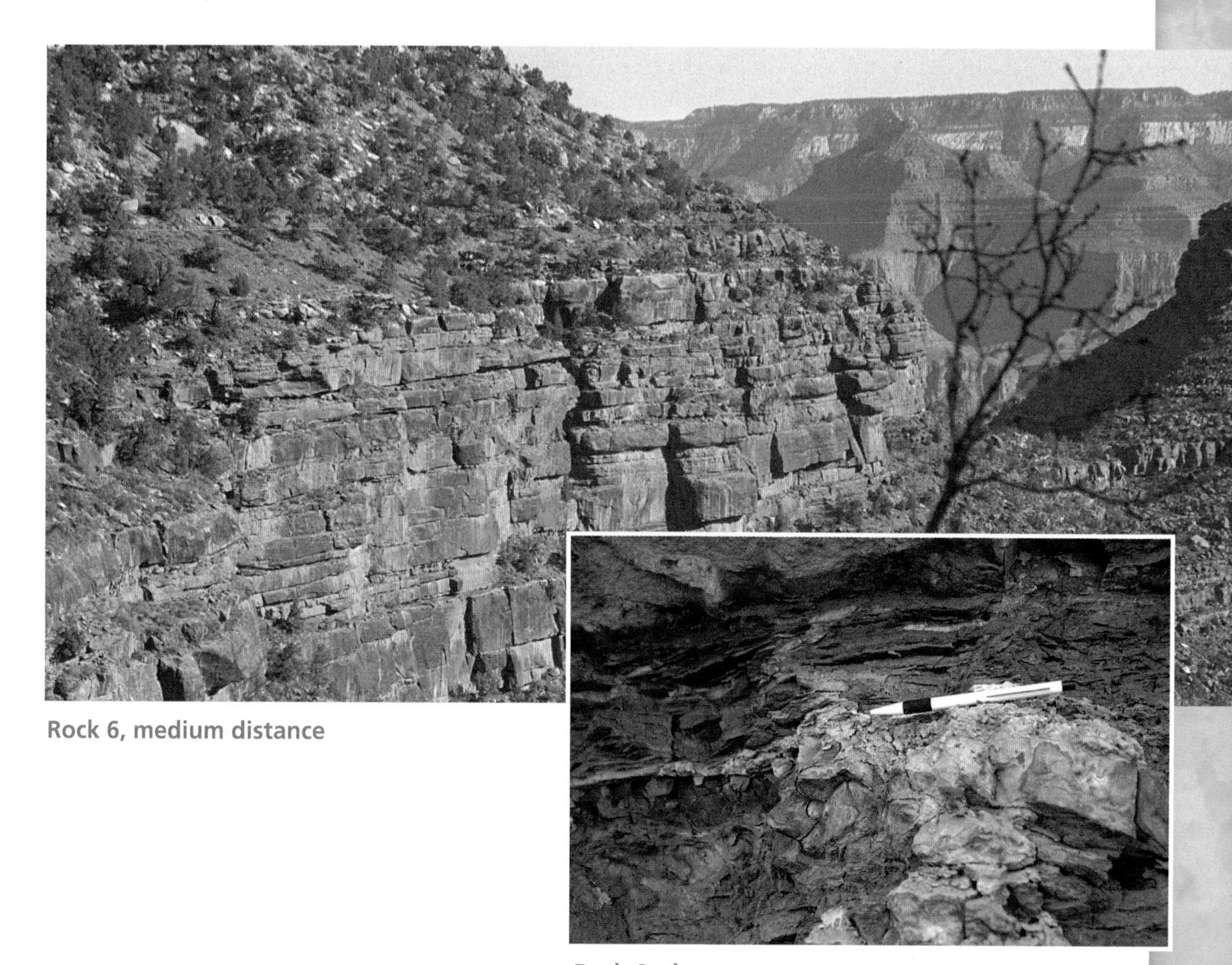

Rock 6, medium distance

Rock 6: close-up

Rock 5, medium distance

Rock 5 close-up

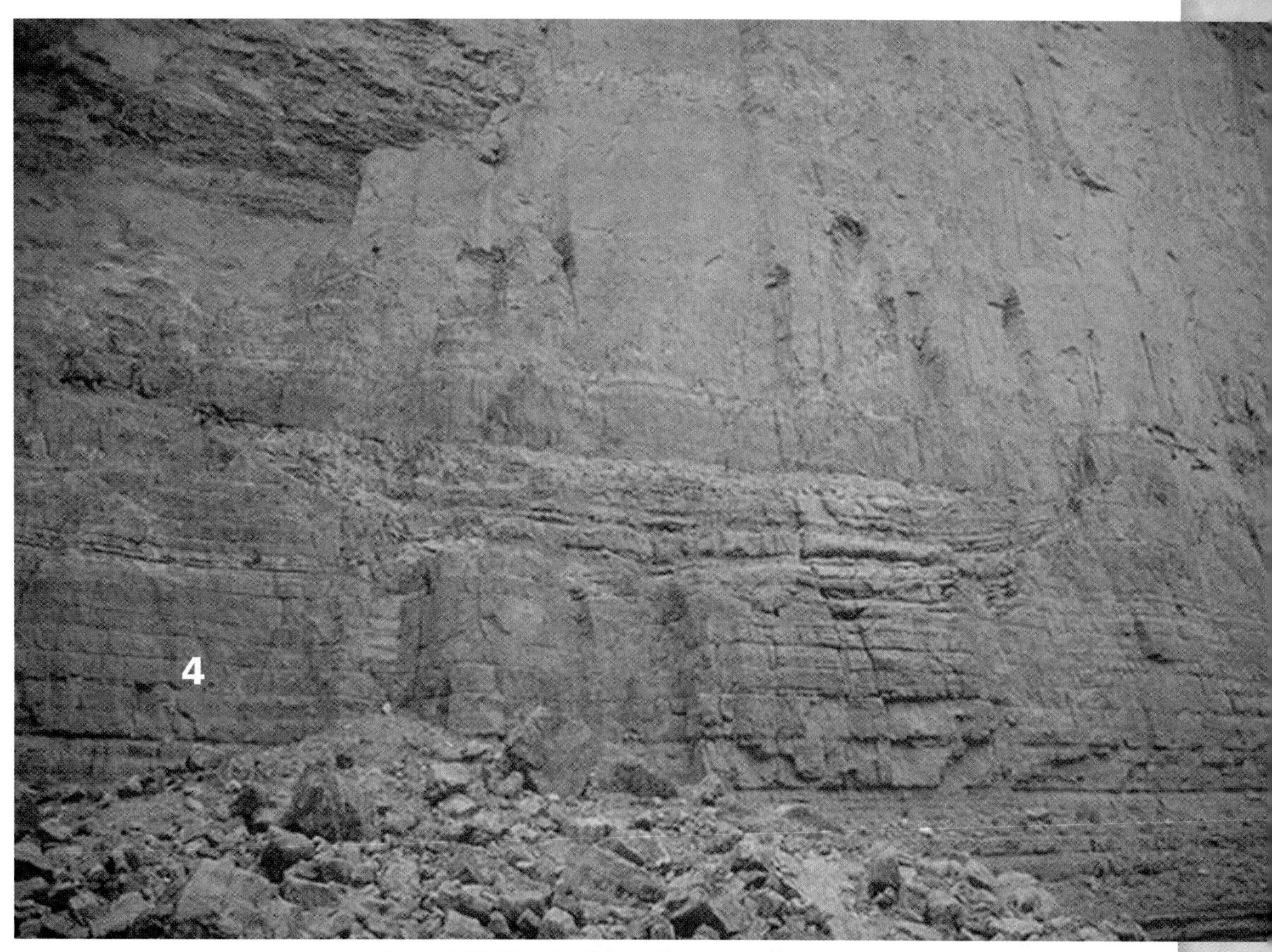

Contact between rocks 4 and 5

Rock 4 close-up

Rock 3, medium distance

Rock 3 close-up

Colorado Plateau Map

Colorado Plateau Rocks

Zion	Canyonlands	Bryce Canyon	Arches	Petrified Forest	Capitol Reef
		Straight Cliffs Formation			
	?1	Tropic Shale	Mancos Shale		Mancos Shale
	?2	Dakota Sandstone	Dakota Sandstone		Dakota Sandstone
	Morrison Formation	Morrison Formation	Morrison Formation		Morrison Formation
	Summerville Formation	Winsor Member	Summerville Formation		Summerville Formation
	Summerville Formation	Winsor Member	Summerville Formation		Summerville Formation
	Curtis Formation	Curtis Formation	?4		Curtis Formation
	Entrada Sandstone	Entrada Sandstone	Entrada Sandstone		Entrada Sandstone
Carmel Formation	Carmel Formation	Carmel Formation	Carmel Formation		Carmel Formation
Navajo Sandstone	Navajo Sandstone	Navajo Sandstone	Navajo Sandstone		Navajo Sandstone
Kayenta Formation	Kayenta Formation		Kayenta Formation		?9
Wingate Sandstone	Wingate Sandstone		Wingate Sandstone		?10
Chinle Formation	?3		Chinle Formation	Chinle Formation	
Moenkopi Formation	Moenkopi Formation		Moenkopi Formation	Moenkopi Formation	
Kaibab Limestone	Cutler Formation		?5	De Chelly Sandstone	
	Cutler Formation			Glorieta Sandstone	
	Cutler Formation			?6	
	Rico Formation			?7	
	Rico Formation			?8	
	Hermosa Group			Supai Group	

Erosion on the Colorado Plateau

Cedar Mountain, Arizona

Monument Valley, Arizona

Arches National Park, Utah

Mexican Hat, Utah

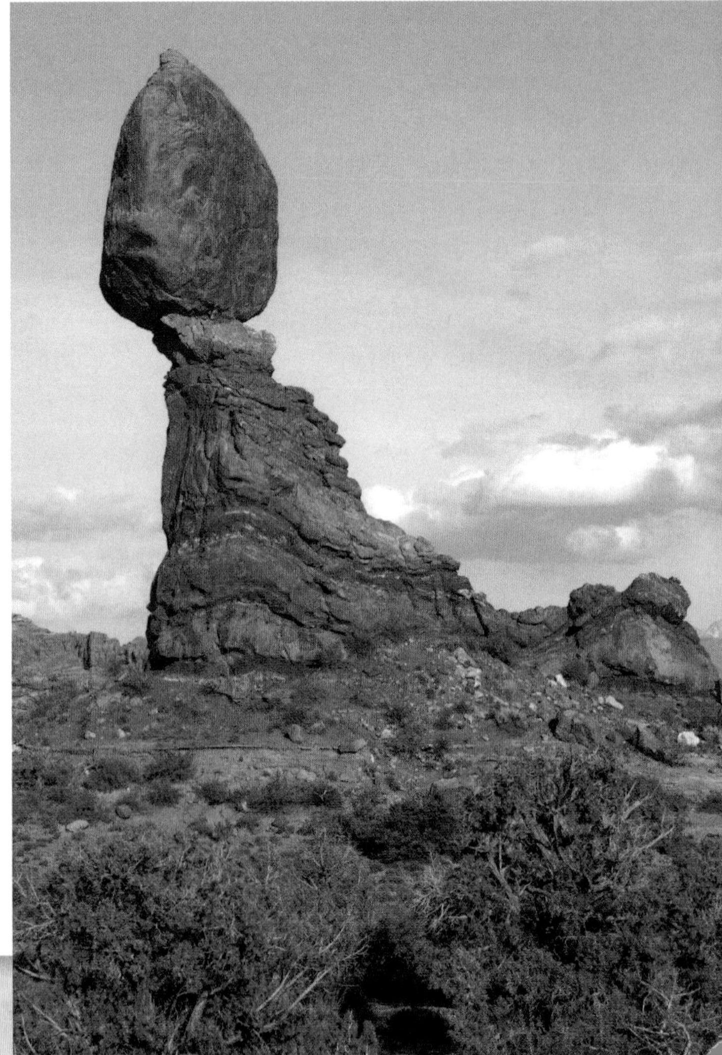
Balanced Rock, Arches National Park, Utah

Bryce Canyon National Park, Utah

Colorado National Monument, Colorado

Canyon de Chelly National Monument, Arizona

Agathla Peak, New Mexico

Goblin Valley State Park, Utah

Wentworth Scale of Rock Particle Sizes

CLASSIFICATION	PARTICLE SIZE (DIAMETER)
Boulder	Above 256 mm
Cobble	64–256 mm
Pebble	4–64 mm
Gravel (or granule)	2–4 mm
Very coarse sand	1–2 mm
Coarse sand	0.5–1 mm
Medium sand	0.25–0.5 mm
Fine sand	0.125–0.25 mm
Very fine sand	0.062–0.125 mm
Silt	0.004–0.062 mm
Clay	Less than 0.004 mm

Sand Analysis

You can observe several properties of sand that provide clues about the sand's origin. These include shape, composition, grain size, and sorting.

Shape

Geologists use different names to describe the shape of sand particles. As sand is moved from one place to another by water, wind, or ice, it gets worn down or abraded. You can identify the shape of a sand grain when you look at it through a magnifier.

Angular has sharp edges. Edges are hardly worn off.
Subangular shows slight abrasion. Corners and edges are worn off slightly.
Subrounded shows more abrasion. Many edges are noticeably worn off.
Rounded has edges smoothed and worn by abrasion. Original shape is still recognizable.
Well-rounded has no edges or corners. They have been totally worn off.

PARTICLE ROUNDNESS

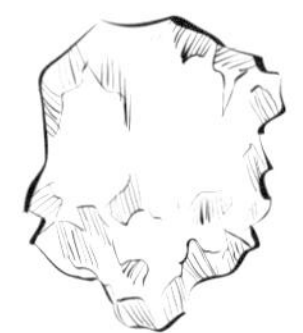

Angular

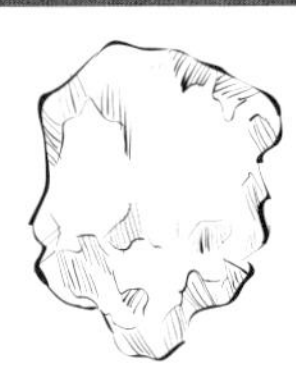
Subangular

Subrounded

Rounded

Well-rounded

Composition

The composition of sand depends on the rock from which it formed. It can help you trace the sand back to its source. Sand found on California beaches is most often composed of quartz with some feldspar. Geologists have identified its source as nearby mountains. The mountains are composed of granites and other igneous rocks—sources of quartz and feldspar. Some beaches in Hawaii are composed of black sand. Black sand has its source in basalt lava flows.

Grain size

Sand is classified into several sizes. See the Wentworth scale for more details about size.

Sorting

Sorting describes how the sand has been separated by size. Sorting depends on how and how far the sand moved. Water does a good job of sorting particles by size and density. Wind sorts particles even more, but only the smaller sizes. Glacial ice does little sorting. Well-sorted sand has particles all the same size; poorly sorted sand includes many sizes.

SORTING

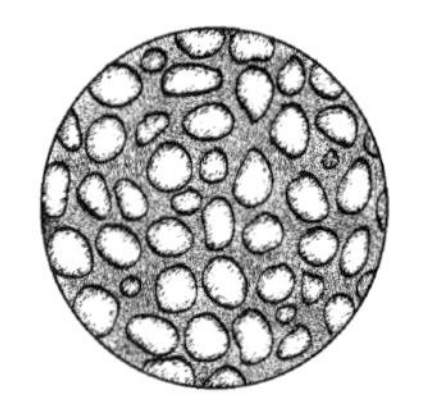
Very well sorted

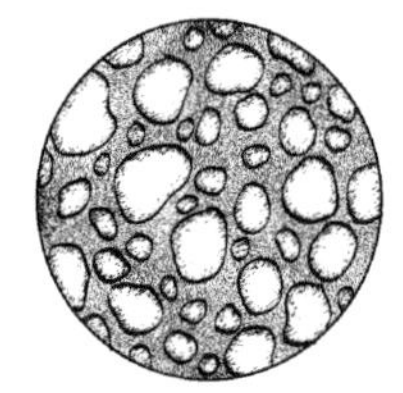
Well sorted

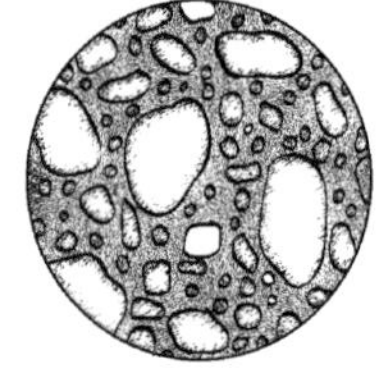
Moderately well sorted

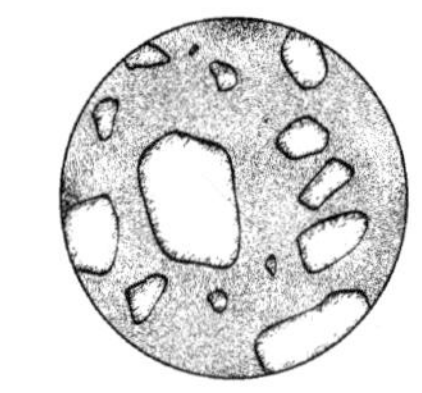
Poorly sorted

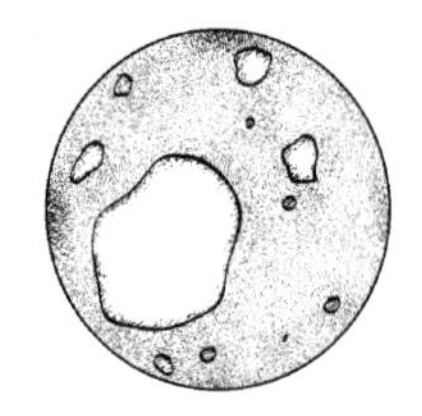
Very poorly sorted

Sand Comparisons

Asilomar State Beach, Monterey Peninsula, California

Green Sand Beach, Big Island Hawaii

Sunset Beach, Hawaii

Mountain Creek, Chilnualna area, Yosemite National Park, California

Sand dunes near Cairo, Egypt

Beach, Amelia Island, Florida

Sand on the Move

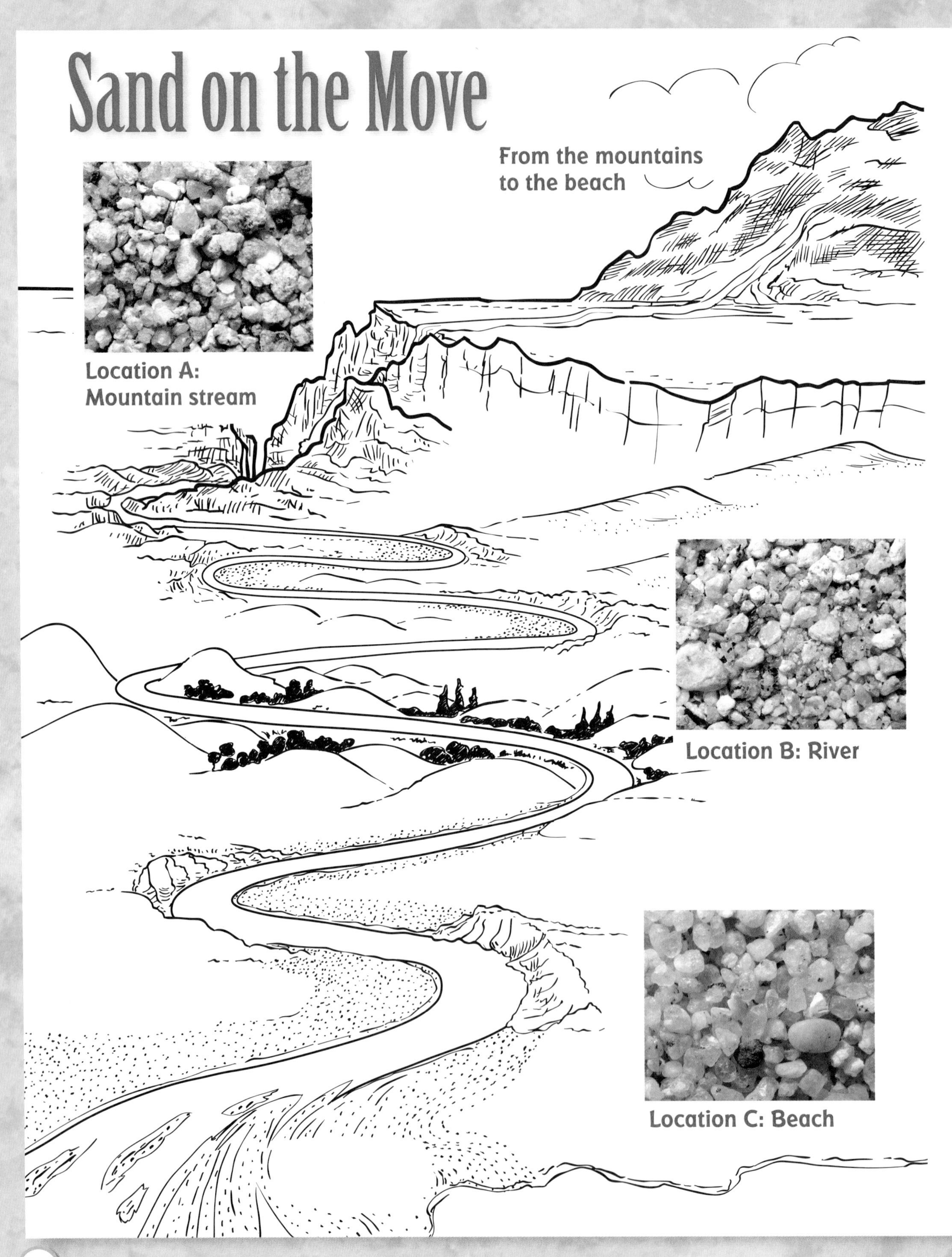

Mystery Sands

Sand 1

Sand 2

Modern Sedimentary Environments

Mudflat/beach

Mudflat

Sandy sea bottom

Sand dune in a mountain valley

Sandy sea bottom

Swamp

River channel and floodplain

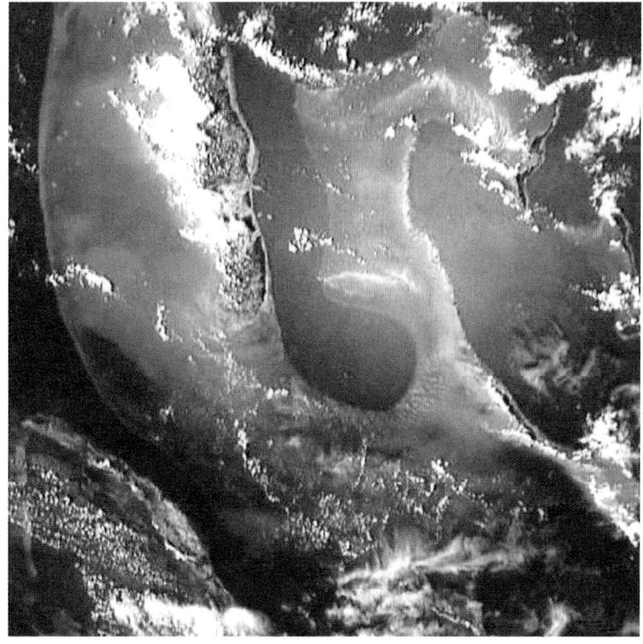

Satellite view of the Bahama Banks

Beach with assorted rocks

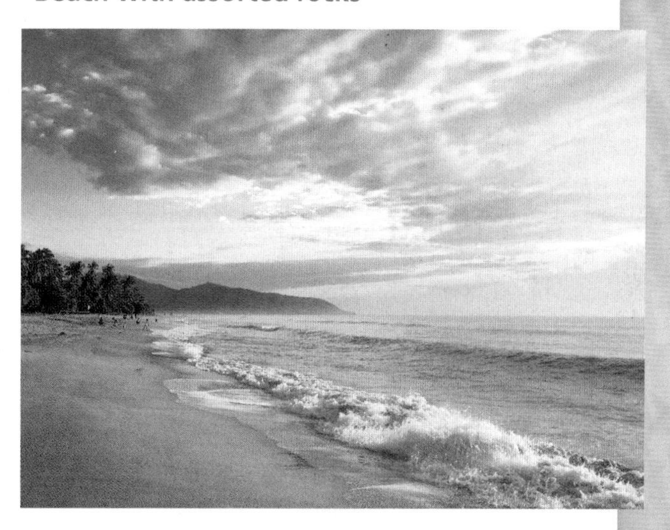

Waves on sandy beach

Features of Sedimentary Rocks

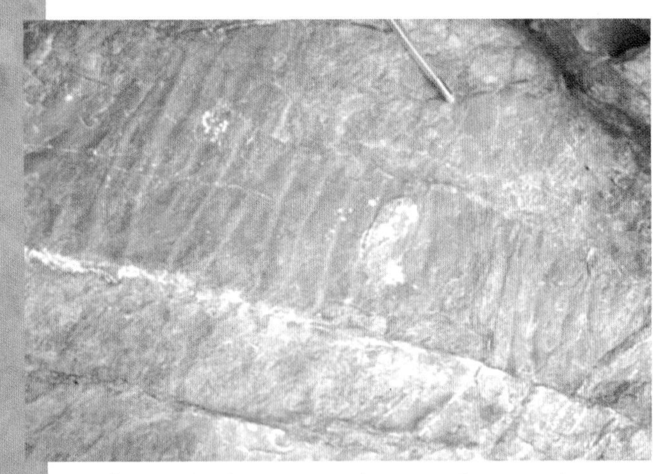

Ripple marks in Mesnard quartzite, northern Michigan

Ripple marks in Hermit shale, northern Arizona

Mudcracks in shale, northern Michigan

Ripple marks in Bright Angel shale, northern Arizona

Crossbeds in sandstone, Arches National Park

Crossbed in Navajo sandstone, Coyote Buttes, Arizona

Fossil Indentification

Invertebrates and Plants

Kingdom: Animalia
Phylum: Mollusca
Class: Bivalves

Modern examples

Clams, oysters, mussels, and scallops

Description

All bivalves are marine and freshwater mollusks with two shells. The shells are held together with a ligament. For most bivalves, such as clams and mussels, the shells are mirror images. Bivalves feed by filtering microscopic organisms from water.

Age of first appearance

Early Cambrian period 535 million years ago (mya)

Kingdom: Animalia
Phylum: Mollusca
Class: Gastropods

Modern examples

Snails, abalones, sea slugs, slugs, limpets, and conches

Description

Gastropods include marine and land-dwelling snails and slugs. Gastropods have one shell.

Age of first appearance

Early Cambrian period 535 mya

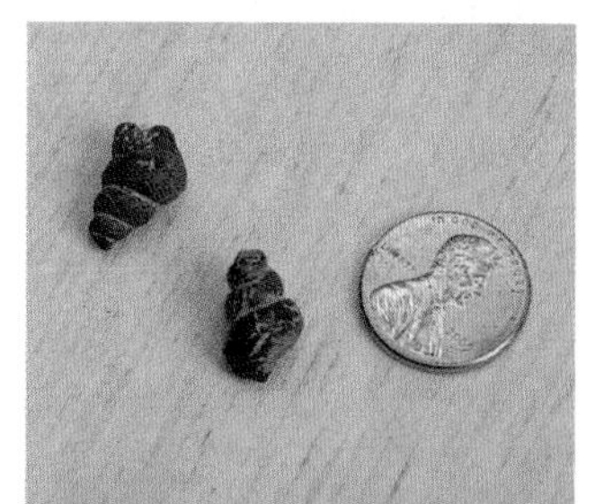

Kingdom: Animalia
Phylum: Mollusca
Class: Cephalopods

Modern examples

Chambered nautiluses, squids, cuttlefish, and octopuses

Description

Many modern cephalopods do not have shells. Those with shells pump air in and out of the shell's chambers to create buoyancy. Cephalopods pump jets of water to move around. Fossil cephalopods have both straight and coiled shells.

Age of first appearance

Early Cambrian period 535 mya

Became dominant during the Ordovician period 488 mya

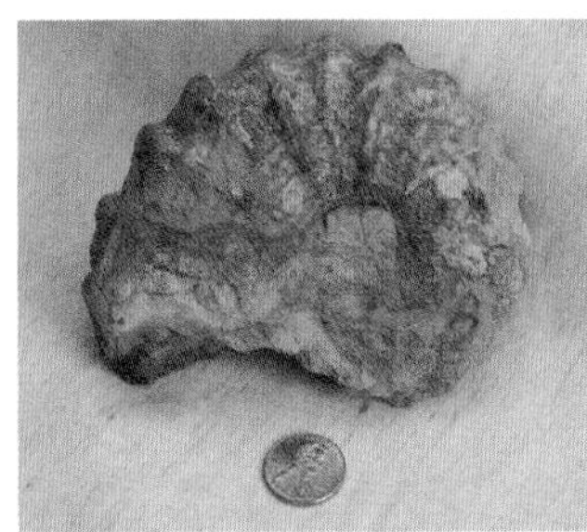

Kingdom: Animalia
Phylum: Brachiopoda

Modern example

Lingula brachiopods

Description

Brachiopods are marine organisms with two shells. The top shell is larger than the bottom shell. There is a small opening at the shell's hinge where a small "foot" (pedicle) comes out to attach to the rocks and hold the brachiopod in place.

Age of first appearance

Early Cambrian period 535 mya

Became most abundant during the Ordovician period 488 mya and again during the Mississippian period 359 mya

Kingdom: Animalia
Phylum: Cnidaria
Class: Anthozoan

Modern example

Corals

Description:

Anthozoans are small marine organisms that secrete the calcium carbonate structures they live in. Modern corals build large colonial structures and are important reef builders.

Age of first appearance

First appeared in the Cambrian period but rare until the Ordovician period 488 mya.

Solitary horn corals were dominant during the Silurian period (443 mya) and extinct by the end of the Permian period (251 mya).

Modern corals became common during the Jurassic period (200 mya).

Kingdom: Animalia
Phylum: Echinodermata
Class: Crinoid and Blastoid

Modern examples

Sea lilies and feather stars

Description

Crinoids and blastoids look like a plant with a long stalk and a feathery head (or calyx). The calyx filters microscopic food from sea water. Blastoids and crinoids existed at the same time. They can be distinguished from each other only if the calyx is present.

Age of first appearance

First appeared in the Ordovician period 488 mya

Mass extinction at the end of the Permian period 251 mya

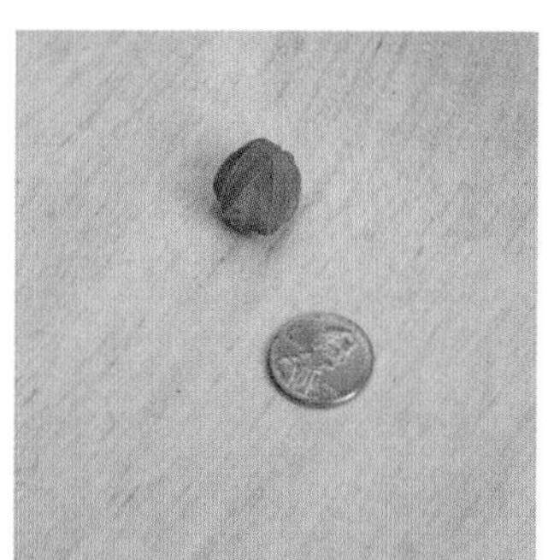

Kingdom: Animalia
Phylum: Echinodermata
Class: Echinoid

Modern examples

Sea urchins, sand dollars, sea biscuits, sea cucumbers, and brittle stars

Description

Echinoids are marine organisms that have a roundish calcium carbonate skeleton that is covered with spines.

Age of first appearance

Upper Ordovician period 450 mya

Most abundant during the Devonian period (416 mya) and almost became extinct at the end of the Permian period (251 mya)

Kingdom: Animalia
Phylum: Bryozoa

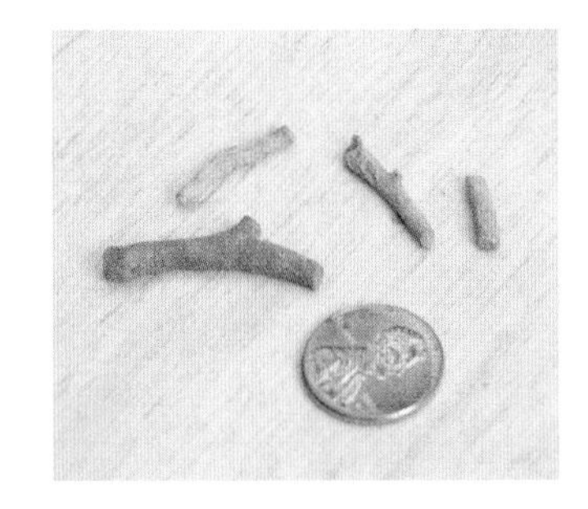

Modern example

Bryozoa

Description

Bryozoa are small marine organisms that form colonies similar to corals. They are filter feeders and live in tropical and temperate waters.

Age of first appearance

Early Ordovician period 480 mya

Kingdom: Animalia
Phylum: Porifera

Modern example

Sponges

Description

Sponges are primarily soft-bodied and rarely preserved. They produce small spines, or spicules, within their bodies. The spicules are usually made of silica (quartz). Sponges do not have a nervous, digestive, or circulatory system.

Age of first appearance

Precambrian era 780 mya

Kingdom: Animalia
Phylum: Arthropoda
Class: Trilobite

Modern examples

Insects, crabs, lobsters, shrimp, and spiders

Description

Trilobites were marine arthropods. Their bodies had three lobes and thrived in the Paleozoic seas until their extinction in the Permian period.

Age of first appearance

Lower Paleozoic era 540 mya

Flourished until extinction at the end of the Permian period (251 mya)

Kingdom: Animalia
Phylum: Chordata
Subphylum: Vertebrata

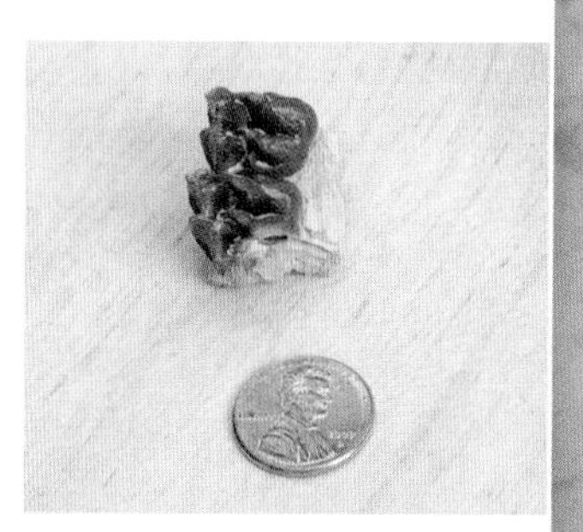

Modern examples

Fish, sharks, reptiles, and mammals

Description

Vertebrata have backbones. The first vertebrates were fish. They flourished during the Devonian period. The largest land vertebrates were the dinosaurs that dominated Earth until their extinction at the end of the Paleozoic era. Modern vertebrates are found on land, in the air, and in the ocean. The largest modern vertebrate is the blue whale.

Age of first appearance

Cambrian period 525 mya

Fish dominated the Devonian period 416 mya.

Dinosaurs dominated the Mesozoic era until extinction at the end of the Cretaceous period 65 mya.

Mammals appeared at the Triassic period 250 mya.

Mammals dominate the Cenozoic era (65 mya to present).

Kingdom: Plantae

Modern examples

Trees, ferns, and cycads

Description

Plants obtain energy from the Sun and produce their own food through photosynthesis. The appearance of land plants made it possible for animals to move from the seas to land in the early Mississippian period (359 mya). Huge coal beds formed during the Pennsylvanian period (318 mya) from lush vegetation that grew on land and in swamps.

The first plants probably evolved from green algae. The early land plants included mosses, ferns, and horsetails. The first trees were conifers and cycads.

Age of first appearance

Ordovician period 488 mya

Kingdom: Protista

Modern examples

Algae, stromatolites, and cyanobacteria

Description

Stromatolites are among some of the oldest fossils found. Modern stromatolites are rare but can be found forming in shallow sections of the ocean.

Age of first appearance

Early Precambrian era 3.2 billion years ago

Grand Canyon Fossils

Kaibab formation

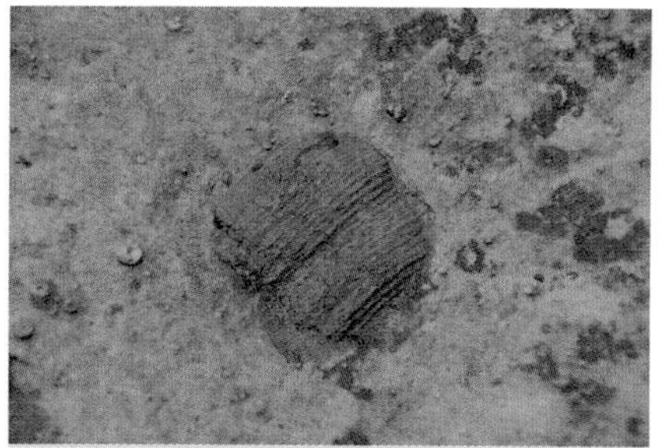

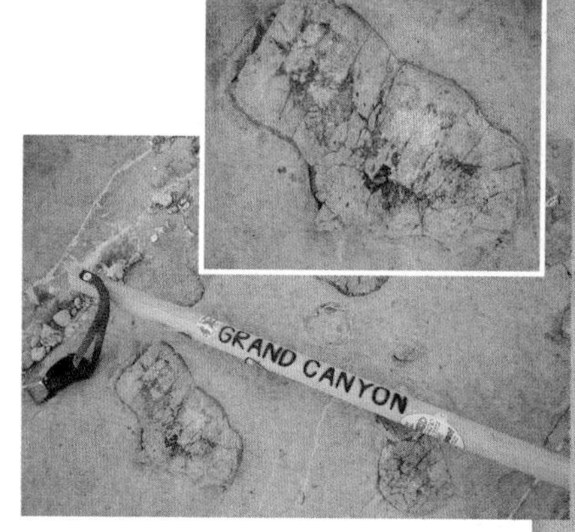

Kaibab fossil

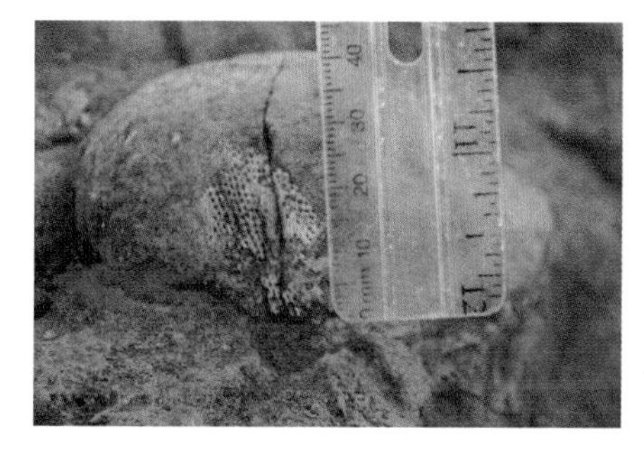

Toroweap limestone

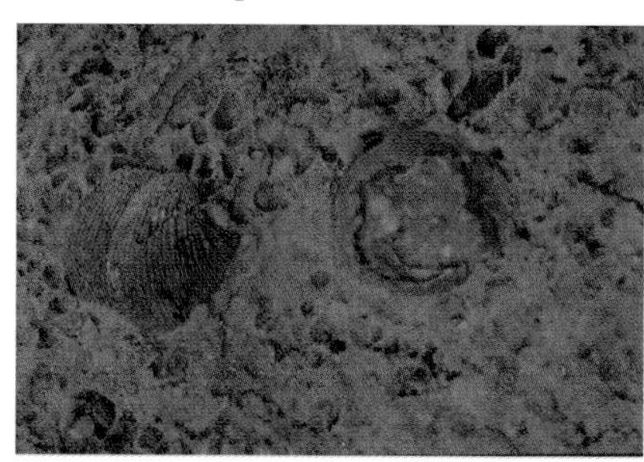

Coconino sandstone

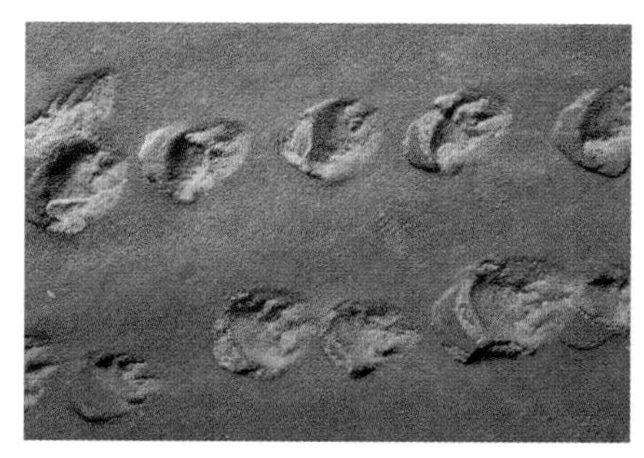

Hermit formation

Supai Group

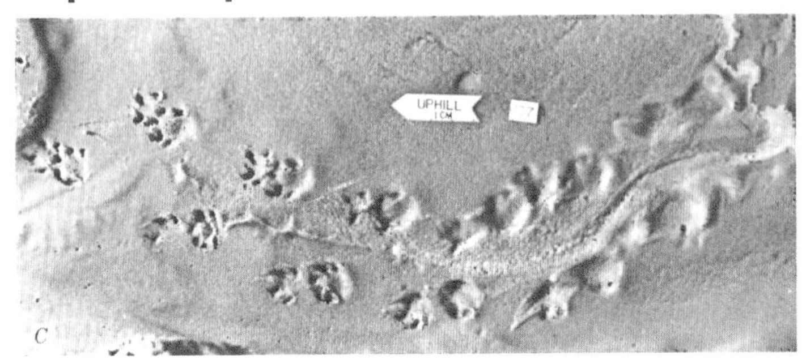

Redwall limestone

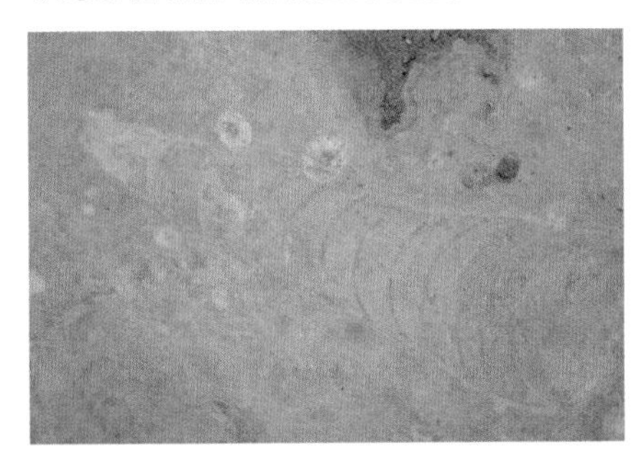

Muav limestone

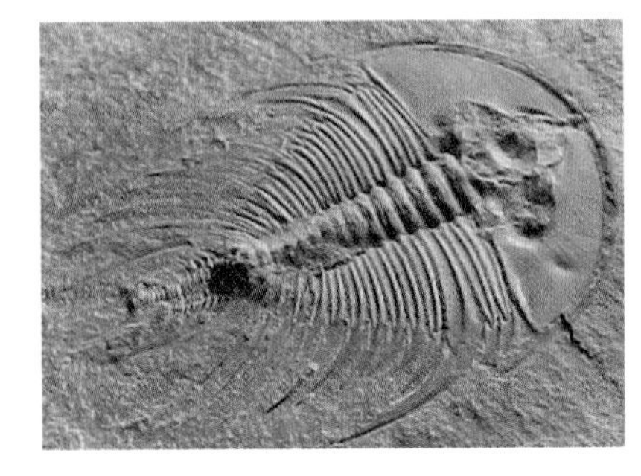

Bright Angel shale

Tapeats sandstone

The Geologic Time Scale

ERA	PERIOD	EPOCH	AGE (MYA)	ORIGIN OF NAME
Cenozoic (Greek for recent life)	Quaternary	Holocene	0.01	Greek for wholly recent
		Pleistocene	2.6	Greek for most recent
	Tertiary	Pliocene	5.3	Greek for more recent
		Miocene	23	Greek for less recent
		Oligocene	34	Greek for slightly recent
		Eocene	56	Greek for dawn of the recent
		Paleocene	66	Greek for early dawn of recent
Mesozoic (Greek for middle life)	Cretaceous		145	Chalk in southern England
	Jurassic		201	Jura Mountains, Switzerland
	Triassic		252	Rocks in Germany (*tri* = three)
Paleozoic (Greek for ancient life)	Permian		299	Province of Perm, Russia
	Pennsylvanian		323	State of Pennsylvania
	Mississippian		359	Mississippi River
	Devonian		419	Devonshire, county of England
	Silurian		444	Silures, Celtic tribe of Wales
	Ordovician		485	Ordovices, Celtic tribe of Wales
	Cambrian		541	Cambria, Roman for Wales
Precambrian			4,600	Before Cambrian

ERAS Geologic eras include major spans of time based on the life-forms that have been found in rocks of that age.

PERIODS The geologic periods have less-clear dividing lines than eras do. It took nearly 100 years to divide the periods. The names and time in the periods are based on outcrops in England, Germany, Russia, Switzerland, and the United States. Some names come from the geographic areas where the rocks appear on the surface (like Jurassic). Other names are based on the characteristics of the rocks (like Cretaceous).

EPOCHS Epochs are subdivisions of the Tertiary and Quaternary periods. English geologist Sir Charles Lyell (1797–1875) came up with these subdivisions. They are based on the percentage of fossils that are represented by animals and plants still living today. The other periods are also divided into epochs. Those epochs are used mainly by geologists who specialize in the rocks of that period.

Index-Fossil Key

BRACHIOPODS

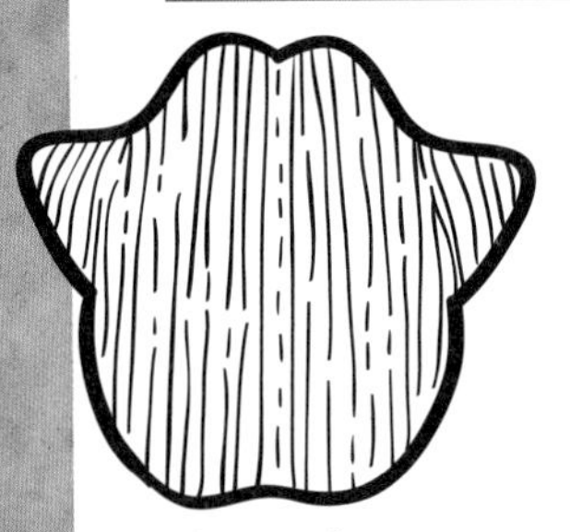

1 cm
Peniculauris
early to middle Permian

1 cm
Composita humulis
early Mississippian

1 cm
Lingula
early Triassic

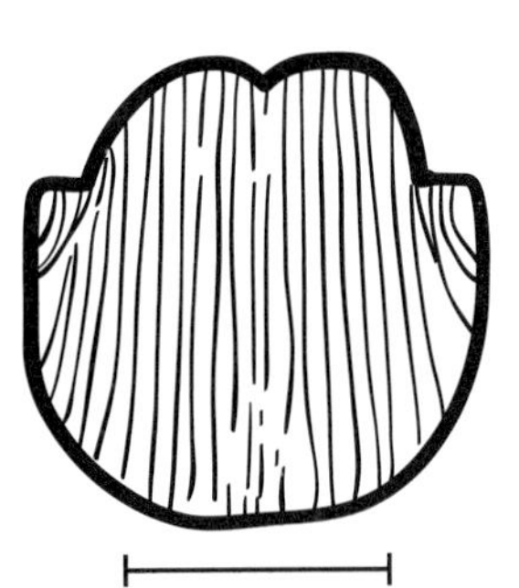

2 cm
Inflatia
late Mississippian

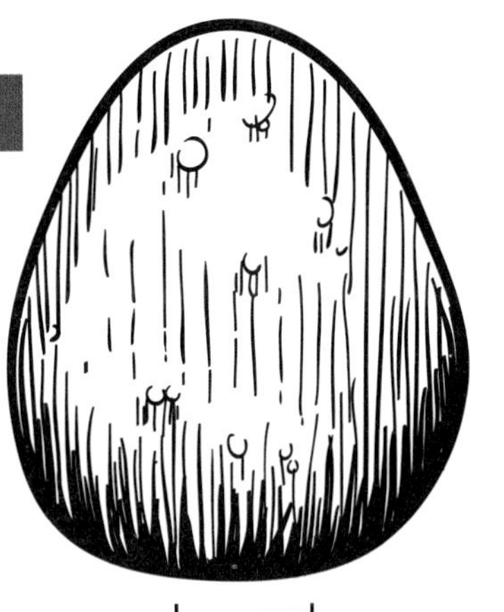

0.5 cm
Ovatia (Linoproductus)
middle Mississippian

1 cm
Rhipidomella
early Mississippian

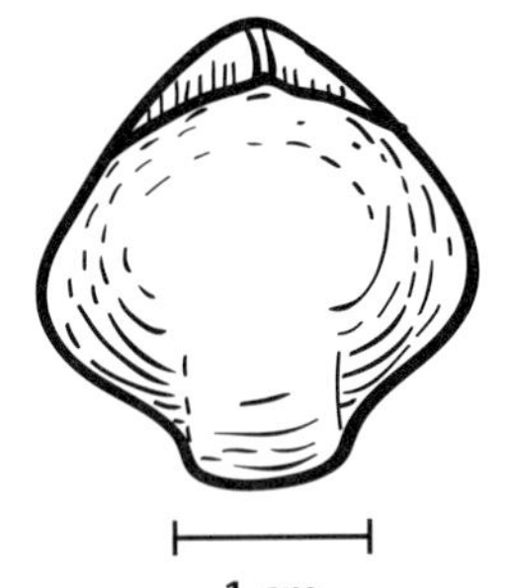

1 cm
Composita trilobata
late Pennsylvanian

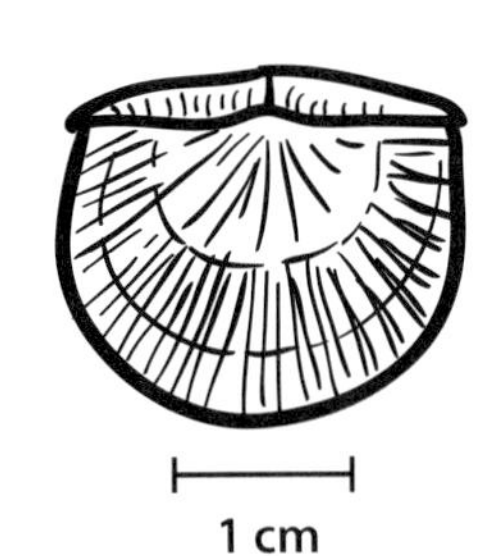

1 cm
Derbyia crassa
late Pennsylvanian

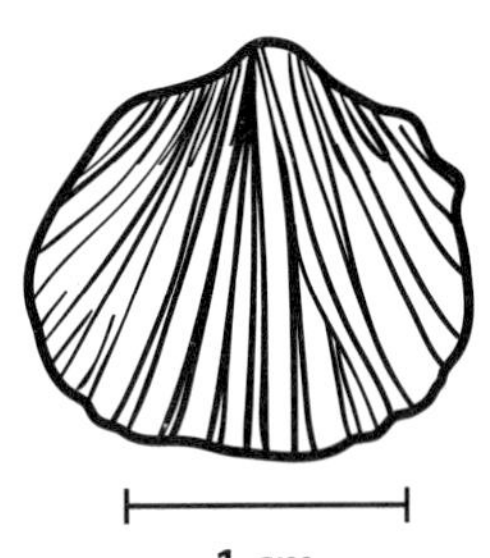

1 cm
Platystrophia
late Ordovician

TRILOBITES

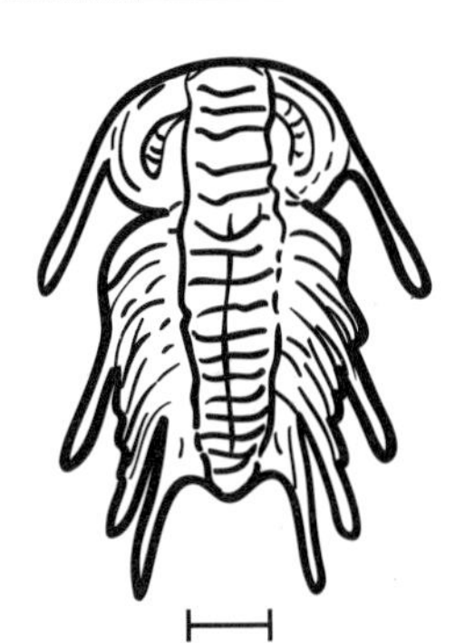

1 cm
Albertella
middle Cambrian

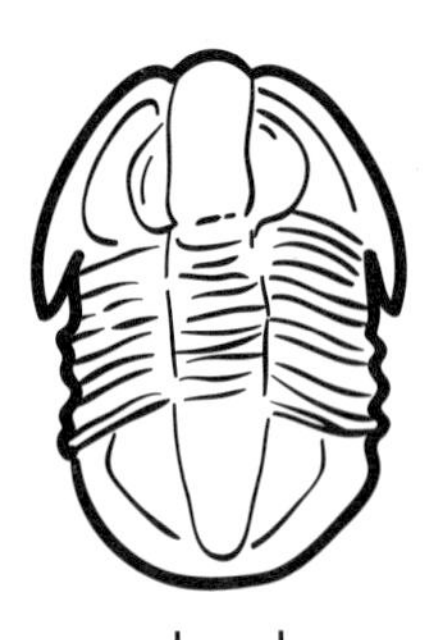

1 cm
Glossopleura
middle Cambrian

1 cm
Olenellus
early Cambrian

CORALS

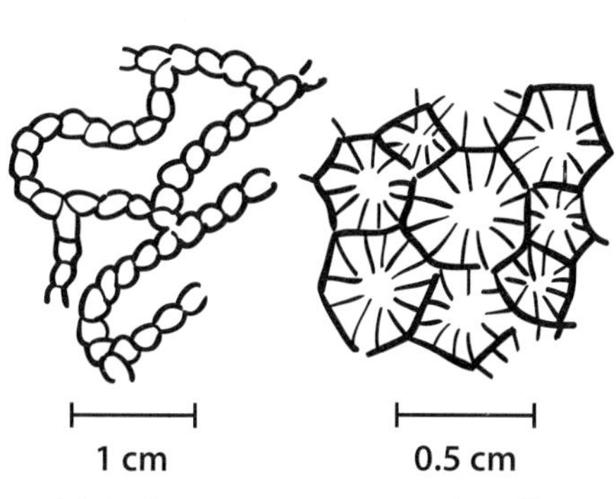

1 cm
Halysites
middle Silurian

0.5 cm
Favistella
late Ordovician

1 cm 0.2 cm
Thamnopora
middle Devonian

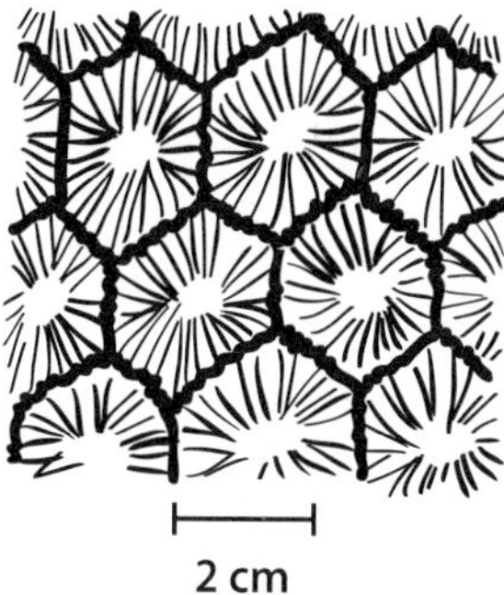

2 cm
Hexagonaria
middle Devonian

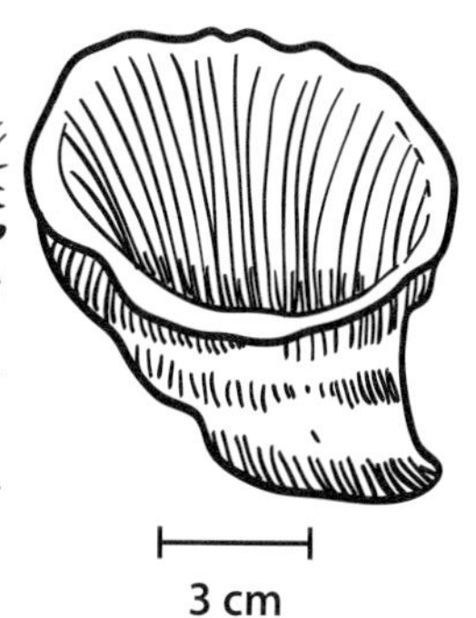

3 cm
Axofistulophyllum
late Silurian

CEPHALOPODS

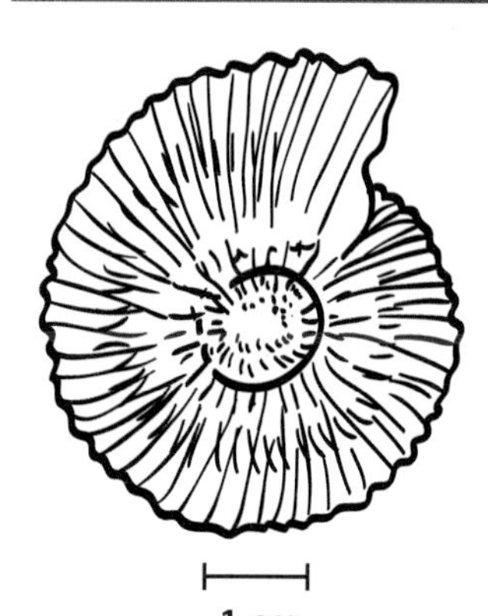

1 cm
Cadoceras
middle to late Jurassic

1 cm
Anasibirites
early Triassic

PLANTS

1 cm
Wingatea
late Triassic

2 cm
Otozamites
late Triassic

SPONGE

1 cm
Actinocoelia
middle Permian

BIVALVE

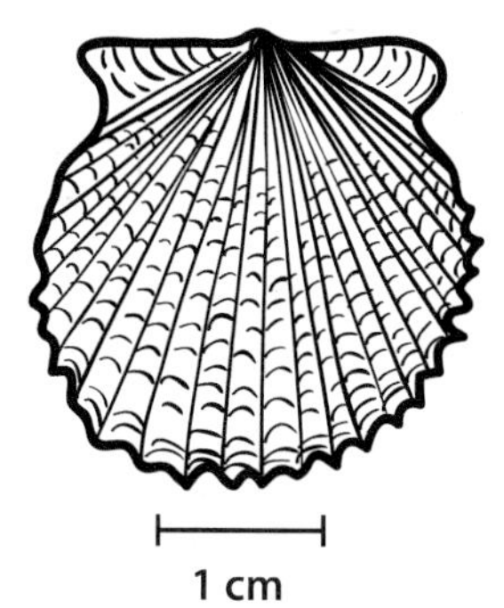

1 cm
Acanthopecten
middle Permian

GASTROPOD

1 cm
Viviparus
late Jurassic

GRAPTOLITE

1 cm
Tetragraptus
early Ordovician

The Great Unconformity

Typical Earth Rocks

Sedimentary Rocks

Composed of broken rock fragments, seashells, and chemicals such as calcium carbonate. Formed from layers of sediment.

ROCK TYPE	DESCRIPTION	IMAGE
Sandstone	Sandstone consists of grains ranging in size from 0.062 to 2 mm in diameter. The grains are usually made of quartz.	
Shale	Shale is composed of clay. Particles are much smaller than those found in sandstone. Most shales are gray, but they can also be red, green, black, or brown. Most can be split into layers.	
Limestone	Limestone can be composed of the remains of sea creatures or from minerals precipitated from water. Calcium carbonate is the main ingredient. Limestone fizzes when acid is dropped on it.	
Conglomerate	Conglomerate is composed of large, rounded pieces of rocks. The grains are larger than those that make up sandstone.	
Breccia	Breccia is composed of large, angular pieces of rocks. The grains are larger than those that make up sandstone.	
Coal	Coal is made from plants living in swampy environments millions of years ago. Color is usually black. Soft coal can be transformed by heat and pressure (metamorphism) into harder grades.	

Igneous Rocks

These rocks form when earth material melts into a liquid, then cools to form solid rock.

	INTRUSIVE ROCKS	EXTRUSIVE ROCKS
QUARTZ-RICH ROCKS	**Granite**  Granite cools from magma in the crust and has large crystals. It makes up a major part of continental crust. Granite is mainly composed of quartz and feldspar with smaller amounts of mica and pyroxene. Granite is commonly used as a building material.	**Rhyolite**  Rhyolite is the extrusive equivalent of granite. It comes from volcanoes that form on continental crust. It cools very quickly and is therefore very fine-grained.
	Diorite 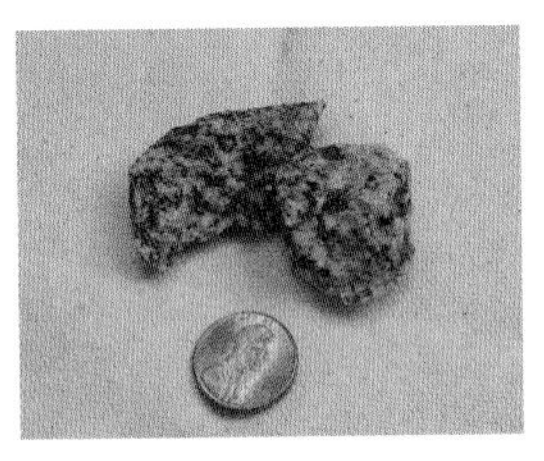 Diorite is an intrusive rock with a mineral composition that is between granite and gabbro.	**Andesite**  Andesite is the extrusive equivalent of diorite and has a mineral composition between rhyolite and basalt.
	**Gabbro**  Gabbro cools from magma in oceanic crust and is found in rift zones. Gabbro is composed of dark, dense minerals such as pyroxene, amphibole, and olivine, as well as plagioclase feldspar.	**Basalt**  Basalt is the extrusive equivalent of gabbro. It is a very dark, dense, fine-grained rock. Oceanic crust is composed of basalt. It is found at the mid-ocean rift zones. The Hawaiian Islands are composed of basalt.

EXTRUSIVE ROCKS (GLASSY)

These rocks cool so quickly that no crystals can form. Their mineral composition depends on where they form.

VARIABLE MINERALS

Obsidian

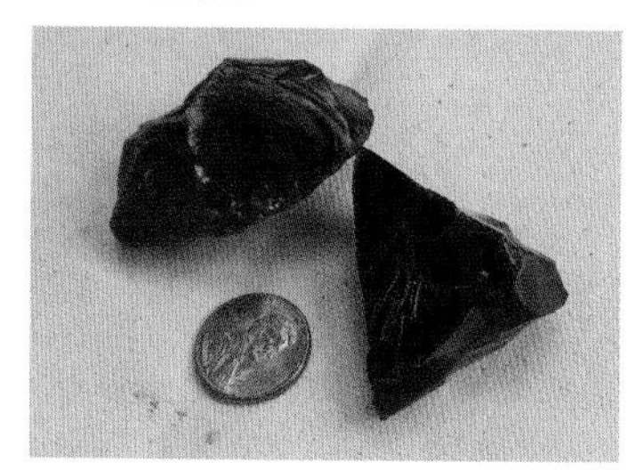

Obsidian is volcanic glass. It is usually dark.

Obsidian was an important rock for ancient people because they could fashion it into very sharp tools.

Pumice

Pumice forms when pressurized lava is ejected so quickly that bubbles form. The bubbles are preserved in the cooled rock.

Pumice is light enough to float on water. Huge rafts of pumice were found floating in the Pacific Ocean 20 years after the eruption of Krakatoa in Indonesia.

Tuff

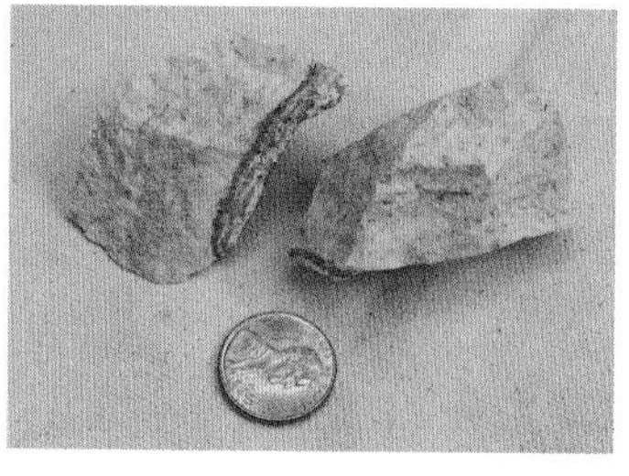

Tuff forms from volcanic ash. The hot ash is ejected from a volcano and welds together before it cools. Tuff could also be considered a sedimentary rock.

The giant moai figures on Easter Island are composed of basaltic tuff.

Scoria

Scoria forms the same way that pumice does. It is not as frothy and has more rock than bubbles. Scoria is often ejected from volcanoes.

Scoria is often used as landscaping rock.

Metamorphic Rocks

Any rock can become metamorphosed. The mineral and chemical composition of a rock determines its origin. The texture of the rock points to the extent of the metamorphism. As metamorphism increases, the texture becomes more fine-grained, layered, and flaky.

Source rock	Less metamorphism	More metaphorphism
Granite Igneous		**Gneiss** Gneiss comes from rock that has been subjected to extreme heat and pressure. The source rock for gneiss is often granite. But it can also be from a sedimentary rock or another metamorphic rock, like schist. When under heat and pressure, the minerals in the rock partially melt and flatten out. Layers form perpendicular to the pressure.
Shale Sedimentary	**Slate** Slate forms when heat and pressure are applied to shale. Slate looks like shale, but slate is much harder because of the compression. To tell the difference between a shale and a slate, tap a hard surface with each rock. The slate will produce a sharper, higher-pitched ping than the shale.	**Schist** Schist is a highly metamorphosed rock that can come from different source rocks, including slate. As metamorphism increases, the rock becomes flaky. Some minerals will melt and form new minerals. Garnets form in metamorphic rocks and are often found in schists.

Metamorphic Rocks *continued*

Source rock	Metamorphism
Limestone  Sedimentary	**Marble** 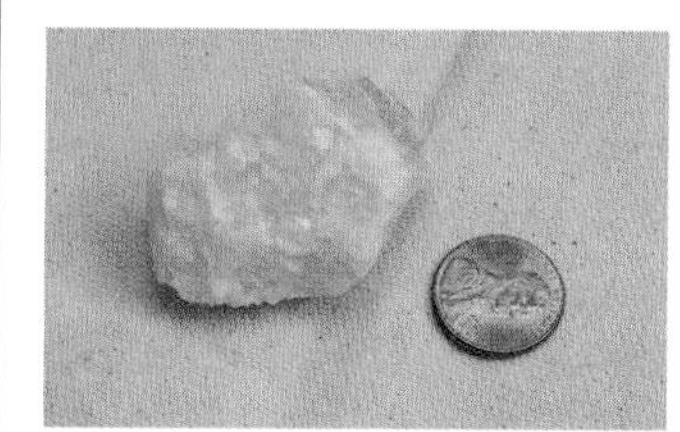 Marble is a limestone that has undergone metamorphism. The heat and pressure causes the calcium carbonate (calcite or $CaCO_3$) to recrystallize. It is harder than limestone. Calcite will fizz with dilute hydrochloric acid. Most sedimentary structures are lost during metamorphism. Some marbles still contain traces of fossils and streaks and swirls from impurities in the original limestone. Marble has always been a popular building material. Because it is made of calcite, it will weather when exposed to acidic rain.
Sandstone  Sedimentary	**Quartzite**  Quartzite forms when sandstone is metamorphosed. As the sand grains are subjected to heat and pressure and partially melt, the pore spaces between the individual grains become filled with quartz. Water easily flows through sandstone but does not flow through quartzite. The color of quartzite is caused by slight impurities found in the original sandstone. Iron oxide makes the rock pink or reddish.

Thin Sections of Igneous Rocks

Photomicrographs

What is a thin section?

A thin section is a piece of rock or mineral sliced so thin that you can easily see through it. Thin sections are used to study how rock interacts with light.

To make a thin section, a lab technician cuts a sample of a rock to fit a glass slide. The technician grinds the sample to 0.03 mm thick. Geologists look at thin sections through microscopes.

What is a photomicrograph?

A photomicrograph is an image taken through a microscope. Photomicrographs of thin sections should include a note about the scale of the image.

Photomicrographs of igneous rocks

The images on this page are photomicrographs of igneous rocks. Compare the three rocks and record your notes in your lab notebook.

- What do you think the different shapes and colors in the thin sections represent?
- What similarities and differences do you observe among the three rocks?
- Compare the photomicrographs to the rock samples that you have. What do you observe?

If you want to learn more about making thin sections, visit FOSSweb.com.

Granite (field of view 8 mm)

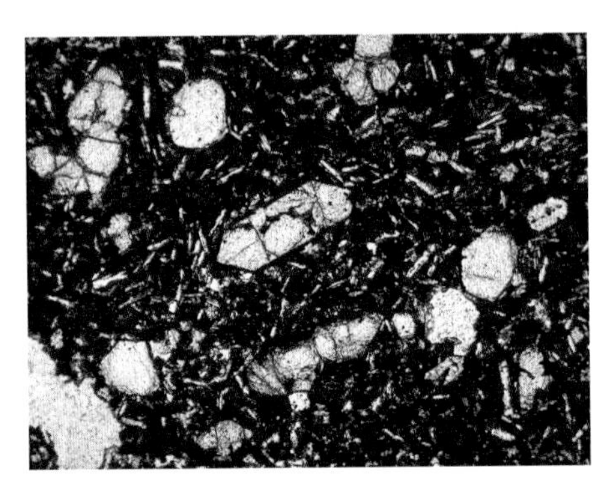

Basalt (field of view 3 mm)

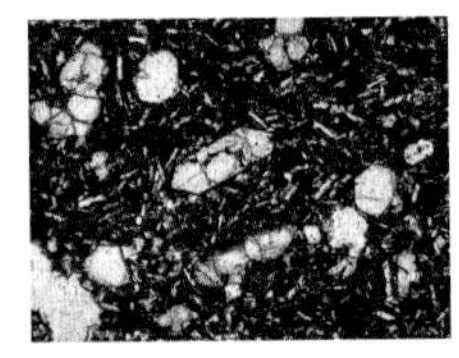

The basalt image at the same scale as the granite photomicrograph

Obsidian (field of view 1.2 mm)

The obsidian image at the same scale as the granite photomicrograph

Map of the Pacific Northwest: Igneous-Rock Locations

Volcanoes

Shield Volcanoes

Shield volcanoes are built almost entirely of lava flows. Lava pours out of a central vent and flows in all directions. The lava builds a broad, gently sloping cone. Some geologists suggest that the profile of these volcanoes looks like a warrior's shield.

Mauna Loa on the Big Island of Hawaii is a shield volcano.

Some of the largest volcanoes in the world are shield volcanoes. The Hawaiian Islands were built mainly by shield volcanoes. Mauna Loa is the world's largest shield volcano, reaching over 8,534 meters (m) from the ocean floor all the way to its top above the water's surface.

Stratovolcanoes or Composite Volcanoes

Stratovolcanoes (also known as composite volcanoes) are large, steep-sided cones. They are built of alternating layers of lava flows, volcanic ash, cinders, and volcanic bombs (large chunks of lava that were blown from the crater and cooled in mid-air). Most stratovolcanoes have a crater at the summit.

Mount Rainier in Washington is a good example of a stratovolcano.

Some of Earth's most beautiful mountains are stratovolcanoes. These include Mount Fuji in Japan, Mount Cotopaxi in Ecuador, Mount Shasta in California, Mount Hood in Oregon, and Mount St. Helens and Mount Rainier in Washington. Stratovolcanoes often erupt explosively and their eruption is a danger to nearby life and property.

Calderas

Calderas form when large, explosive volcanic eruptions shoot out many cubic kilometers of magma. When the magma is removed from beneath the volcano, the volcano collapses into the empty space. The huge depression that forms is a caldera. Some calderas are more than 25 kilometers (km) in diameter and several kilometers deep.

Crater Lake in Oregon is a caldera. Crater Lake is the deepest lake in the United States (592 m deep) and one of the ten deepest lakes in the world.

Somma Volcanoes

A somma volcano is a caldera that has been partially filled by a new central cone. The name comes from Monte Somma, a stratovolcano in southern Italy. It has a summit caldera that the cone of Vesuvius holds.

Vesuvius is the stratovolcano in the center of this image. The larger rim north of Vesuvius is the wall of the Monte Somma caldera, which formed around 17,000 years ago. Vesuvius's latest eruption was in 1944.

Cinder Cones

Cinder cones are built from particles and blobs of lava blown out of a single vent. As the blobs fly through the air, they break into small fragments that solidify and fall as cinders around the vent. They form a circular or oval cone. Most cinder cones have a bowl-shaped crater at the top. They are relatively short, usually no taller than 300 m.

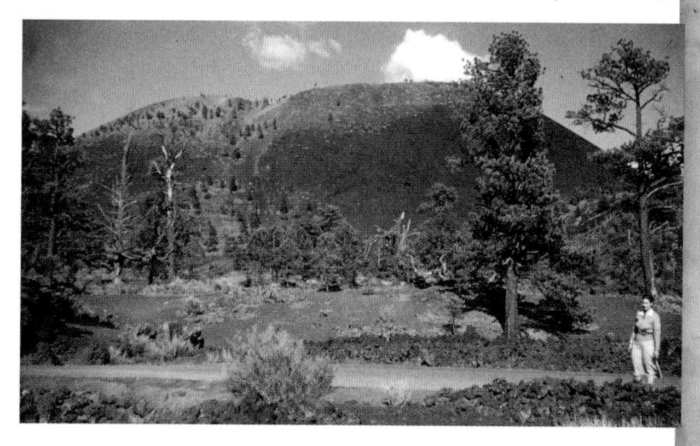

Sunset Crater in Arizona is a cinder cone. It erupted more than 900 years ago.

Lava Plateaus

Sometimes lava pours out quietly from long fissures instead of central vents. The lava floods the surrounding countryside time after time, forming broad lava plateaus. The Columbia Plateau in Washington and Oregon is a lava plateau.

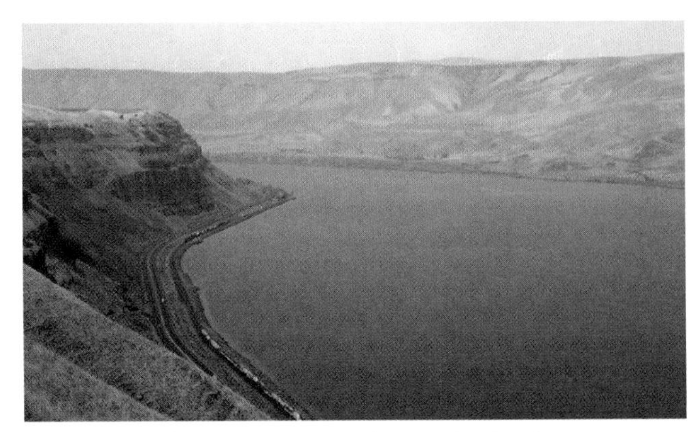

The Columbia River cuts through the lava layers of the Columbia Plateau between Washington and Oregon.

Lava Domes

Lava that is too sticky to flow far from its vent forms steep-sided mounds called lava domes. The igneous rock that forms a lava dome is usually rhyolite or dacite.

Lava domes have formed several times in the crater of Mount St. Helens since its 1980 eruption.

Maars or Tuff Cones

Maars (or tuff cones or rings) are shallow, flat-floored craters. They range in size from 60 to 2,000 m across and from 10 to almost 200 m deep. Most are filled with water to form natural lakes. They have low rims made of loose fragments of volcanic rocks and rocks torn from the walls of a volcanic vent or pipe below. The ash that erupts from the vent eventually solidifies into a rock called tuff.

Fort Rock in Oregon is a tuff ring.

Submarine Volcanoes

Magma can erupt through fractures in ocean crust. Submarine volcanoes can form from the erupting lava, which cools and creates basalt. These volcanoes can grow into islands, with time and plenty of magma.

The pillow-like lava in this image is erupting from the submarine volcano Loihi. Loihi Seamount is forming off the coast of Hawaii's Big Island.

Tuyas

A tuya is a volcano under a glacier. Tuyas are common in Iceland. They can cause huge floods called *jökulhlaups* when they erupt and melt the glacial ice.

A glacier covers the caldera of this tuya in Iceland.

Complex or Compound Volcanoes

A complex volcano or compound volcano has more than one feature. These volcanoes erupt in different ways during their history. They may include stratovolcanoes, cinder cones, calderas, and other volcanic features.

Yellowstone is a complex volcano, well known for its lava falls, geysers, hot springs, calderas, and cinder cones.

Grand Canyon Rock Columns

Layered Paleozoic Rocks
1. Kaibab Formation
2. Toroweap Formation
3. Coconino Sandstone
4. Hermit Shale
5. Supai Group
6. Surprise Canyon Formation
7. Redwall Limestone
8. Temple Butte Formation
9. Muav Limestone
10. Bright Angel Shale
11. Tapeats Sandstone

Grand Canyon Supergroup Rocks
12. Sixtymile Formation
13. Chuar Group
14. Nankoweap Formation
15. Unkar Group

Vishnu Basement Rocks
16. Schists
17. Granites
18. Elves Chasm Gneiss

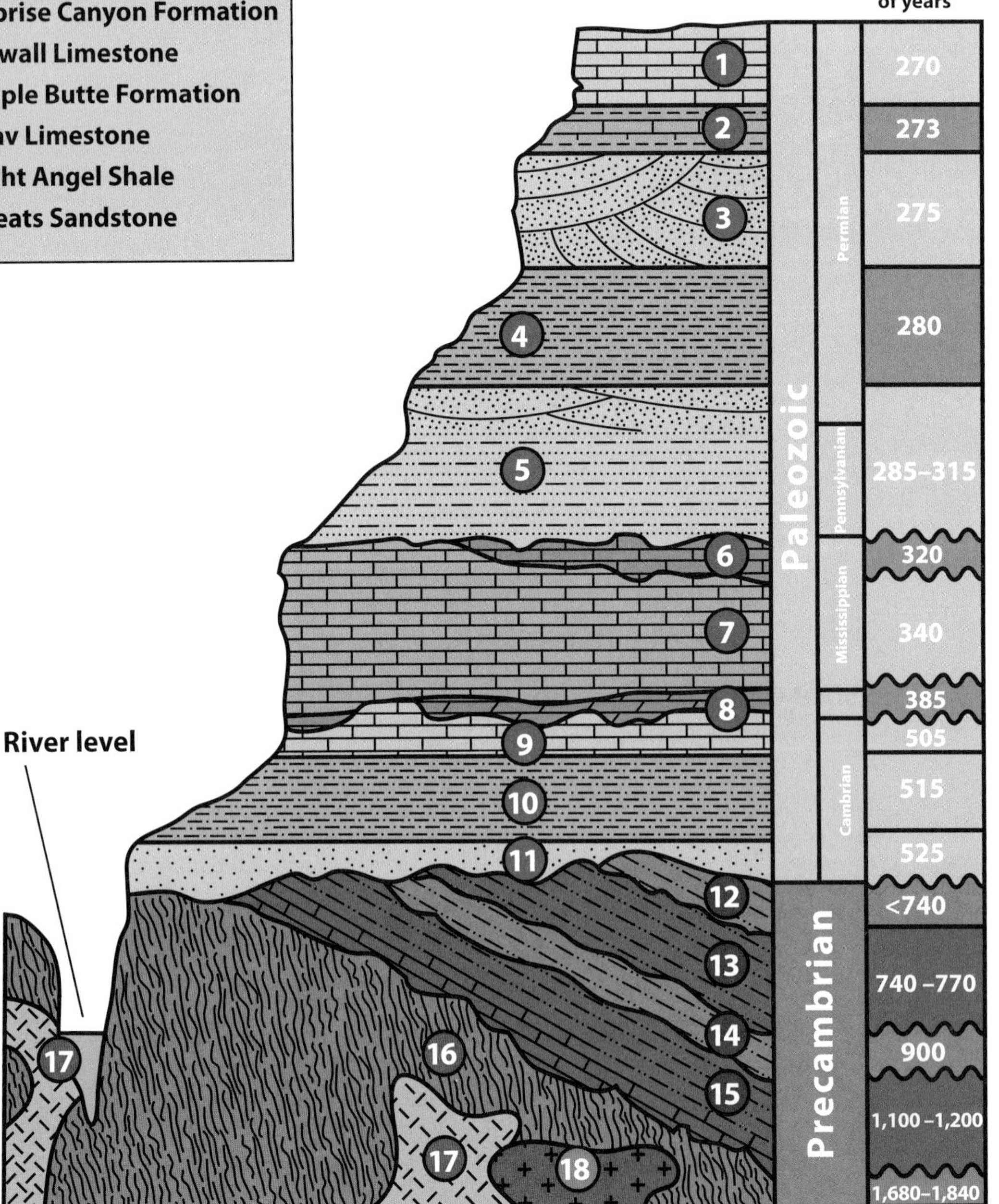

The Other Grand Canyon Rocks

At first glance, it might look like the only rocks in the walls of the Grand Canyon are sedimentary. But if you look closer, you will find that is not quite true.

A few kilometers above Bright Angel is a dark, narrow canyon with walls made of black rocks. There are no layers at the river's edge. Sometimes areas of pink, crystal-filled rock appears. The pink rock looks like it had squeezed into the black rock. In several places, the Tapeats Sandstone rests on top of the uneven surface of these black rocks. The Bright Angel Shale is above the Tapeats Sandstone.

Farther downstream near Lava Falls are more dark rocks. These rocks look like they flow over the edge of the canyon, covering some sedimentary rocks. Some people have described them as "frozen waterfalls" of rock. These rocks flow over a cliff formed by rocks in the Supai Formation.

Dark, flowing rocks
Kaibab Formation
Toroweap Formation
Coconino Sandstone
Hermit Shale
Supai Group
Surprise Canyon Formation
Redwall Limestone
Temple Butte Formation
Muav Limestone
Bright Angel Shale
Tapeats Sandstone
Dark rock with some pink, crystalline rock

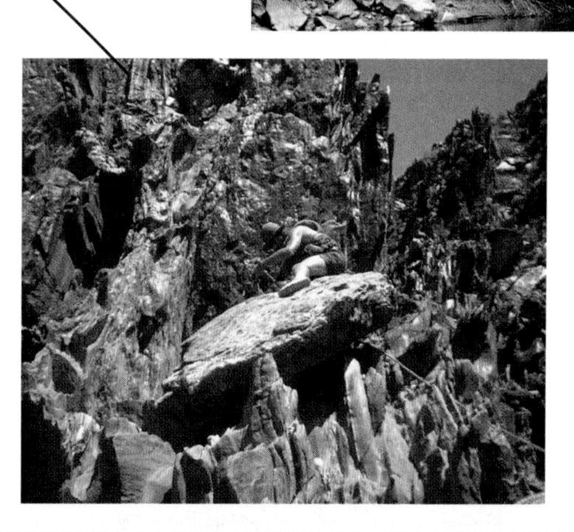

Science Safety Rules

1. Always follow the safety procedures outlined by your teacher. Follow directions, and ask questions if you're unsure of what to do.
2. Never put any material in your mouth. Do not taste any material or chemical unless your teacher specifically tells you to do so.
3. Do not smell any unknown material. If your teacher asks you to smell a material, wave a hand over it to bring the scent toward your nose.
4. Avoid touching your face, mouth, ears, eyes, or nose while working with chemicals, plants, or animals. Tell your teacher if you have any allergies.
5. Always wash your hands with soap and warm water immediately after using chemicals (including common chemicals, such as salt and dyes) and handling natural materials or organisms.

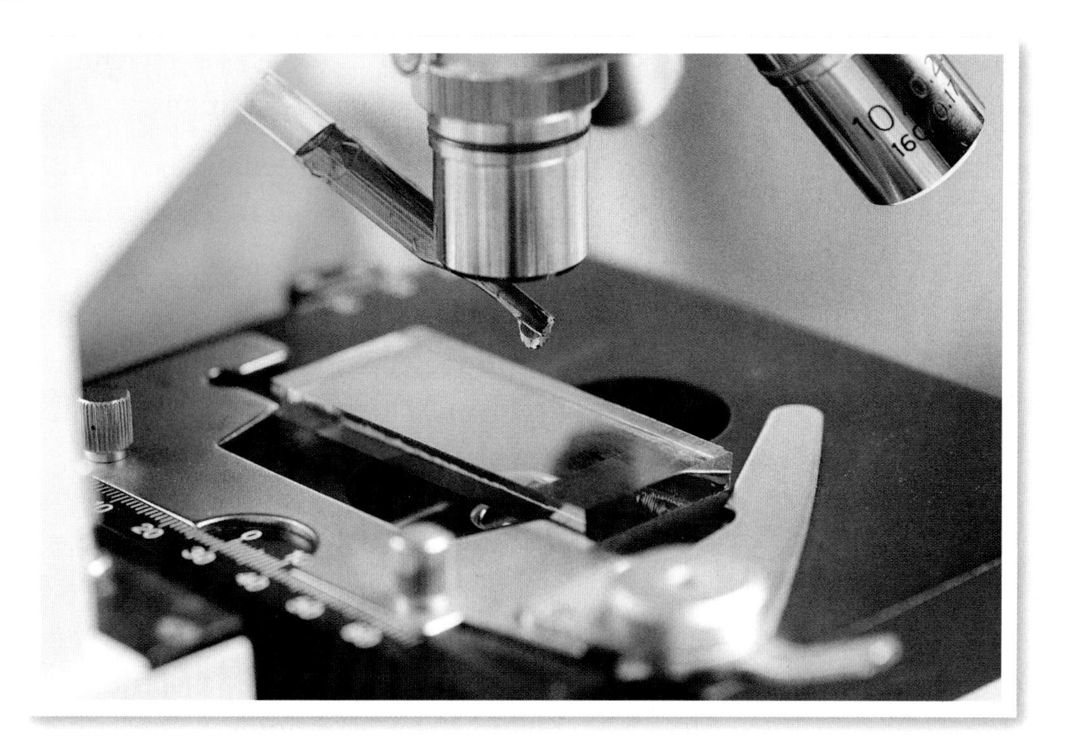

6. Do not mix unknown chemicals just to see what might happen.
7. Always wear safety goggles when working with liquids, chemicals, and sharp or pointed tools. Tell your teacher if you wear contact lenses.
8. Clean up spills immediately. Report all spills, accidents, and injuries to your teacher.
9. Treat animals with respect, caution, and consideration.
10. Never use the mirror of a microscope to reflect direct sunlight. The bright light can cause permanent eye damage.

Glossary

abrasion the grinding and bumping of rocks that cause physical weathering

asthenosphere the fluid portion of the mantle that has special properties

atmosphere the air that surrounds Earth

atom the smallest particle of an element

basin a low-lying area where water may accumulate

bedrock the rock that forms Earth's crust

biosphere all the living things on Earth

calcite a carbonate mineral with the chemical formula $CaCO_3$

caldera a volcanic landform created when part of the land collapses

carbonate a substance made of carbon and oxygen (CO_3)

Cenozoic an era when mammals exist; means "recent animal life"

chemical reaction the process in which two or more substances combine to make one or more new substances that have different properties from the original ones

chemical weathering the process by which rocks are weathered due to reactions with substances in water and air

coal a rock that can be burned to release stored energy

convection the heat transfer in a fluid in which hot fluid rises and cold fluid sinks, setting up a cycle

converge to come together

correlation the process of matching up rock layers from different locations

crossbedding patterns in rock resulting from sand dunes that were blown into patterns by the wind

cross section the surface that is revealed when something is cut through, typically at a right angle to its axis

crude oil unprocessed oil, also called petroleum

crust the rigid outer layer of Earth

crystal an orderly arrangement of atoms and molecules with a specific shape determined by its chemical composition

deposition the settling of sediments

differential erosion the different rates at which layers of weathered earth materials are carried away by water, wind, or ice

Earth-imaging satellite a device launched into space to orbit Earth to record images of its surface, such as GeoEye-1

earthquake a sudden movement of plates within Earth's crust

ecosystem a community of organisms interacting with each other and with the nonliving environment

eon the largest division of geologic time

epoch a subdivision of a geologic period

era a broad span of geologic time based on the general type of life existing during that time

erosion the carrying away of weathered earth materials by water, wind, or ice

erratic a rock that is different from the type of rock found in its current location

extrusive formed on the exterior of Earth

fault a crack in Earth's crust where movement can occur

flood a large amount of water flowing over land that is usually dry

fold a bending of rock layers caused by pressure

folding the deformation of rock layers in Earth's crust as tectonic plates move

foliation to form into thin, leaflike layers

formation rock layers with distinguishable characteristics and sequences of sediments

fossil any remains, trace, or imprint of animal or plant life preserved in Earth's crust

fossil fuel fuel formed by natural processes over long periods of time

fossil record all the fossils on Earth

fumarole a hole or vent near a volcano from which steam rises

geologic time the period of time ranging from the formation of Earth about 4.6 billion years ago to today

geologist a scientist who studies Earth

geosphere the solid rocky surface and the interior of Earth

geyser a hot spring that traps steam in underground spaces and builds pressure, causing periodic eruptions

glacial till sediment from glaciers that is unsorted

glacier a giant body of ice that slowly flows

greenhouse gas a gas that absorbs and radiates heat energy in the atmosphere, effectively trapping heat in the atmosphere

groundwater water stored below Earth's surface

hoodoo a rock shaped like a mushroom or statue that forms when weak rocks erode away and leave stronger rocks behind

horizon a layer in a soil profile

hotspot a volcanic area that forms as a tectonic plate moves over a place heated from deep within Earth

hot spring a naturally occurring warm body of water heated from underground

humus (HEW•mus) bits of dead plant and animal parts in soil

hydrosphere Earth's water, both in the seas and on land

igneous rock a rock that forms when melted rock (magma or lava) hardens

index fossil a fossil that characterizes a particular period of time

infer to form a conclusion from evidence or known facts

interglacial the time between glaciation

intrusive formed in the interior of Earth

karst topography dramatic landforms created by chemical weathering of rocks

landform a natural feature of Earth's surface with a characteristic shape

lava molten rock on the surface of Earth

law of fossil succession a principle that says the same kinds of fossils found in rocks from different places are the same age and that the kinds of plants and animals found as fossils change over geologic time

limestone a sedimentary rock made mostly of calcite

lithosphere the hard outer layer of Earth made of crust and hard, upper mantle; broken into tectonic plates

magma molten rock under the surface of Earth

mantle the layer of Earth below the crust; upper part is solid and lower part is semisolid

Mesozoic an era when dinosaurs existed; means "middle animal life"

metamorphic rock a rock that has changed from another rock because of heat, pressure, or a chemical reaction

mineral a naturally occurring chemical compound

model a representation of something else

molecule a particle made of two or more atoms that are held together with strong bonds

mud pots fumaroles with little water, so mud forms and bubbles at the entrance

oil a substance formed within Earth that can be burned to release stored energy

ooze in geology, a deposit of soft mud in water

paleontology the study of fossils

Paleozoic an era when trilobites, corals, brachiopods, early fish, and early amphibians existed; means "ancient animal life"

peat plant material that has partially decomposed

pedologist a scientist who studies soil

period a span of time within an era

physical weathering the process by which rocks are broken down by breaking and banging

plate a section of Earth's lithosphere

plateau a large area of flat-lying sedimentary rock that has been lifted high above its original elevation

plate boundary the area along an edge of Earth's plates

plate tectonics a theory that says Earth's outer layers are made of moving plates

precipitate to form a solid, insoluble product during a chemical reaction

principle of original horizontality a theory that says sedimentary layers of rock are flat and horizontal

principle of superposition a theory that says sedimentary rocks on the bottom are older than rocks on the top

relative time scale sequence of events based on data that shows the order in which they occur

renewable something that can be replenished

rock cycle transformation that changes one type of rock into another

rock fall a mass of rocks falling to the ground, similar to an avalanche

sandstone a sedimentary rock made of sand particles stuck together

satellite an object that orbits a planet or other object

sediment material that is deposited by water, wind, or ice

sedimentary rock a rock that forms when layers of sediments get stuck together

seismologist a scientist who studies earthquakes

shale a sedimentary rock made of clay or silt

slab pull when solid, dense lithosphere sinks into the softer, less dense asthenosphere, causing the entire tectonic plate to move

soil a mix of humus, sand, silt, clay, gravel, and/or pebbles

soil profile a vertical cross section of soil that shows its many layers or horizons

source rock a rock at the start of a process

stalactite a calcium carbonate deposit shaped like an icicle, hanging from the roof of a cave

stalagmite a calcium carbonate deposit shaped like an upside-down icicle, formed on the floor of a cave

strain a force or pressure that tends to change the shape of a rock

stratigraphy the study of the order and correlation of Earth's rocks

subduction when one tectonic plate slides under another

subduction zone an area where two plates meet and one moves under the other

supereruption the massive eruption of a supervolcano

talus jagged rocks found at the base of cliffs

tectonic the movement of Earth's crust

tectonic plate sections of the lithosphere that move on top of the fluid asthenosphere; composed of Earth's crust and the hard top layer of the mantle

terminal moraine an area of glacial sediment that forms at the farthest reaching point of a glacier

terrain the physical features of land

transform a fault where two plates slide past each other

trench deep areas of the crust where oceanic plates are converging and sinking

unconformity a gap in the geologic record during which either no rocks were deposited or existing rocks eroded

uniformitarianism a theory that says geologic processes observed in the past are the same as those observed today

uplift when sections of Earth's crust rise as tectonic plates move

volcano an opening in Earth's crust where lava, cinders, ash, and gases come to the surface

volocanologist a scientist that studies volcanoes

weathering the process by which larger rocks crack and break apart over time to form smaller rocks

Index